W9-DDY-200

PRAISE FOR *BECOMING A DATA HEAD*

Big Data, Data Science, Machine Learning, Artificial Intelligence, Neural Networks, Deep Learning . . . It can be buzzword bingo, but make no mistake, everything is becoming "data-fied" and an understanding of data problems and the data science toolset is becoming a requirement for every business person. Alex and Jordan have put together a must read whether you are just starting your journey or already in the thick of it. They made this complex space simple by breaking down the "data process" into understandable patterns and using everyday examples and events over our history to make the concepts relatable.

—**Milen Mahadevan**, President of 84.51°

What I love about this book is its remarkable breadth of topics covered, while maintaining a healthy depth in the content presented for each topic. I believe in the pedagogical concept of "Talking the Walk," which means being able to explain the hard stuff in terms that broad audiences can grasp. Too many data science books are either too specialized in taking you down the deep paths of mathematics and coding ("Walking the Walk") or too shallow in over-hyping the content with a plethora of shallow buzzwords ("Talking the Talk"). You can take a great walk down the pathways of the data field in Alex and Jordan's without fear of falling off the path. The journey and destination are well worth the trip, and the talk.

—**Kirk Borne**, Data Scientist,
Top Worldwide Influencer in Data Science

The most clear, concise, and practical characterization of working in corporate analytics that I've seen. If you want to be a killer analyst and ask the right questions, this is for you.

—**Kristen Kehrer**, Data Moves Me, LLC,
LinkedIn Top Voices in Data Science & Analytics

THE book that business and technology leaders need to read to fully understand the potential, power, AND limitations of data science.

—**Jennifer L. L. Morgan**, PhD, Analytical Chemist at
Procter and Gamble

You've heard it before: "We need to be doing more machine learning. Why aren't we doing more sophisticated data science work?" Data science isn't the magic unicorn that will solve all of your company's problems. *Data Head* brings this idea to life by highlighting when data science is (and isn't) the right approach and the common pitfalls to watch out for, explaining it all in a way that a data novice can understand. This book will be my new "pocket reference" when communicating complicated concepts to non-technically trained leaders.

—**Sandy Steiger**, Director, Center for Analytics and
Data Science at Miami University

Individuals and organizations want to be data driven. They say they are data driven. *Becoming a Data Head* shows them how to actually become data driven, without the assumption of a statistics or data background. This book is for anyone, or any organization, asking how to bring a data mindset to the whole company, not just those trained in the space.

—**Eric Weber**, Head of Experimentation & Metrics Research, Yelp

What is keeping data science from reaching its true potential? It is not slow algorithms, lack of data, lack of computing power, or even lack of data scientists. *Becoming a Data Head* tackles the biggest impediment to data science success—the communication gap between the data scientist and the executive. Gutman and Goldmeier provide creative explanations of data science techniques and how they are used with clear everyday relatable examples. Managers and executives, and anyone wanting to better understand data science will learn a lot from this book. Likewise, data scientists who find it challenging to explain what they are doing will also find great value in *Becoming a Data Head.*

—**Jeffrey D. Camm**, PhD, Center for Analytics Impact,
Wake Forest University

Becoming a Data Head raises the level of education and knowledge in an industry desperate for clarity in thinking. A must read for those working with and within the growing field of data science and analytics.

—**Dr. Stephen Chambal**, VP for Corporate Growth at
Perduco (DoD Analytics Company)

Gutman and Goldmeier filter through much of the noise to break down complex data and statistical concepts we hear today into basic examples and analogies that stick. *Becoming a Data Head* has enabled me to translate my team's data needs into more tangible business requirements that make sense for our organization. A great read if you want to communicate your data more effectively to drive your business and data science team forward!

—**Justin Maurer**, Engineering and Data Science Manager at Google

As an aerospace engineer with nearly 15 years experience, *Becoming a Data Head* made me aware of not only what I personally want to learn about data science, but also what I need to know professionally to operate in a data-rich environment. This book further discusses how to filter through often overused terms like artificial intelligence. This is a book for every mid-level program manager learning how to navigate the inevitable future of data science.

—**Josh Keener**, Aerospace Engineer and Program Manager

A must read for an in-depth understanding of data science for senior executives.

—**Cade Saie**, PhD, Chief Data Officer

Gutman and Goldmeier offer practical advice for asking the right questions, challenging assumptions, and avoiding common pitfalls. They strike a nice balance between thoroughly explaining concepts of data science while not getting lost in the weeds. This book is a useful addition to the toolbox of any analyst, data scientist, manager, executive, or anyone else who wants to become more comfortable with data science.

—**Jeff Bialac**, Senior Supply Chain Analyst at Kroger

Gutman and Goldmeier have written a book that is as useful for applied statisticians and data scientists as it is for business leaders and technical professionals. In demystifying these complex statistical topics, they have also created a common language that bridges the longstanding communication divide that has — until now — separated data work from business value.

—**Kathleen Maley**, Chief Analytics Officer at datazuum

Becoming a Data Head

Becoming a Data Head

How to Think, Speak, and Understand Data Science, Statistics, and Machine Learning

ALEX J. GUTMAN

JORDAN GOLDMEIER

WILEY

For my children Allie, William, and Ellen.

Allie was three when she discovered dad was a "doctor."
Puzzled, she looked at me and said, "But, you don't help people. . . ."
In that spirit, I also dedicate this book to you, the reader.

I hope this helps you.
—Alex

For Stephen and Melissa
—Jordan

About the Authors

Alex J. Gutman is a data scientist, corporate trainer, Fulbright Specialist grant recipient, and Accredited Professional Statistician® who enjoys teaching a wide variety of data science topics to technical and non-technical audiences. He earned his Ph.D. in applied math from the Air Force Institute of Technology where he currently serves as an adjunct professor.

Jordan Goldmeier is an internationally recognized analytics professional and data visualization expert, author, and speaker. A former chief operations officer at Excel.TV, he has spent years in the data training trenches. He is the author of *Advanced Excel Essentials* and *Dashboards for Excel*. His work has been cited by and quoted in the Associated Press, Bloomberg Business-Week, and American Express OPEN Forum. He is currently an Excel MVP Award holder, an achievement he's held for six years, allowing him to provide feedback and direction to Microsoft product teams. He once used Excel to save the Air Force $60 million. He is also a volunteer Emergency Medical Technician.

About the Technical Editors

William A. Brenneman is a Research Fellow and the Global Statistics Discipline Leader at Procter & Gamble in the Data and Modeling Sciences Department and an Adjunct Professor of Practice at Georgia Tech in the Stewart School of Industrial and Systems Engineering. Since joining P&G, he has worked on a wide range of projects that deal with statistics applications in his areas of expertise: design and analysis of experiments, robust parameter design, reliability engineering, statistical process control, computer experiments, machine learning, and statistical thinking. He was also instrumental in the development of an in-house statistics curriculum. He received a Ph.D. in Statistics from the University of Michigan, an MS in Mathematics from the University of Iowa, and a BA in Mathematics and Secondary Education from Tabor College. William is a Fellow in both the American Statistical Association (ASA) and the American Society for Quality (ASQ). He has served as ASQ Statistics Division Chair, ASA Quality and Productivity Section Chair, and as Associate Editor for *Technometrics*. William also has seven years of experience as an educator at the high school and college level.

Jennifer Stirrup is the Founder and CEO of Data Relish, a UK-based AI and Business Intelligence leadership boutique consultancy delivering data strategy and business-focused solutions. Jen is a recognized leading authority in AI and Business Intelligence Leadership, a Fortune 100 global speaker, and has been named as one of the Top 50 Global Data Visionaries, one of the Top Data Scientists to follow on Twitter, and one of the most influential Top 50 Women in Technology worldwide.

Jen has clients in 24 countries on 5 continents, and she holds postgraduate degrees in AI and Cognitive Science. Jen has authored books on data and artificial intelligence and has been featured on CBS Interactive and the BBC as well as other well-known podcasts, such as Digital Disrupted, Run As Radio, and her own Make Your Data Work webinar series.

Jen has also given keynotes for colleges and universities, as well as donated her expertise to charities and non-profits as a Non-Executive Director. All of Jen's keynotes are based on her 20+ years of global experience, dedication, and hard work.

Acknowledgments

I've noticed a trend in acknowledgment sections—the author's spouse is often mentioned at the end. I suppose it's a saving-the-best-for-last gesture, but I promised my wife if I ever wrote a book, I'd mention her first to make it perfectly clear whose contributions mattered most to me. So, to my wife Erin, thank you for your love, encouragement, and smile. As I write this, you are taking our three young children on a bike ride, giving me time to write one final page. (I assure all readers this act is a representative sample of our lives this past year.)

I'd also like to thank my parents, Ed and Nancy, for being the best cheerleaders in whatever I do and for showing me what being a good parent looks like, and to my siblings Ryan, Ross, and Erin for their support.

This book is the culmination of many discussions with friends and colleagues, ranging from whether I should attempt to write a book about data literacy to potential topics that should appear in it. Thank you especially to Altynbek Ismailov, Andy Neumeier, Bradley Boehmke, Brandon Greenwell, Brent Russell, Cade Saie, Caleb Goodreau, Carl Parson, Daniel Uppenkamp, Douglas Clarke, Greg Anderson, Jason Freels, Joel Chaney, Joseph Keller, Justin Maurer, Nathan Swigart, Phil Hartke, Samuel Reed, Shawn Schneider, Stephen Ferro, and Zachary Allen.

I'm also indebted to the hundreds of engineers, business professionals, and data scientists I've interacted with, personally or online, who've taught me how to be a better data scientist and communicator. And to my "students" (colleagues) who have given candid feedback about the courses I've taught, I heard you and I thank you.

I'm fortunate to have many academic and professional mentors who've given me numerous opportunities to find my voice and confidence as a statistician, data scientist, and trainer. Thank you to Jeffery Weir, John Tudorovic, K. T. Arasu, Raymond Hill, Rob Baker, Scott Crawford, Stephen Chambal, Tony White, and William Brenneman (who kindly served as a technical editor on this book). It's impossible not to become wiser hanging around a group like that.

Thanks to the team at Wiley: Jim Minatel for believing in the project and giving us a chance, Pete Gaughan and John Sleeva for guiding us through the process, and the production staff at Wiley for meticulously combing through our chapters. And to our technical editors, William Brenneman and Jen Stirrup, we appreciate your suggestions and expertise. The book is better because of you.

Special thanks to my coauthor Jordan Goldmeier, for one obvious reason (the book in your hands) and one not so obvious. Early in my career, I complained to Jordan that people didn't share my interest in statistics and statistical thinking. He said if I'm bothered by it, then it's my obligation to change it. I've been working to fulfill that obligation ever since.

Finally, I'd like to thank my wife Erin one final time (because you've got to save the best for last).

—Alex

I would like to acknowledge the many people who brought this book together.

First, and foremost, I would like to acknowledge my coauthor-in-crime, Alex Gutman. For years, we discussed writing a book together. When the moment was right, we pulled the trigger. I couldn't have asked for a better coauthor.

Thanks to the wonderful folks at Wiley who helped put this together, including acquisition editor Jim Minatel, and project editor John Sleeva. Also, I would like to acknowledge our technical editors, William Brenneman and Jen Stirrup for your hard work reviewing the book. We took your comments to heart.

Last but not least, thank you to my partner, Katie Gray, who always believed in this project—and me.

—Jordan

Contents

CHAPTER 14

CHAPTER 15

Foreword

*B*ecoming a Data Head is well-timed for the current state of data and analytics within organizations. Let's quickly review some recent history. A few leading companies have made effective use of data and analytics to guide their decisions and actions for several decades, starting in the 1970s. But most ignored this important resource, or left it hiding in back rooms with little visibility or importance.

But in the early to mid-2000s this situation began to change, and companies began to get excited about the potential for data and analytics to transform their business situations. By the early 2010s, the excitement began to shift toward "big data," which originally came from Internet companies but began to pop up across sophisticated economies. To deal with the increased volume and complexity of data, the "data scientist" role arose with companies—again, first in Silicon Valley, but then everywhere.

However, just as firms were beginning to adjust to big data, the emphasis shifted again—around about 2015 to 2018 in many firms—to a renewed focus on artificial intelligence. Collecting, storing, and analyzing big data gave way to machine learning, natural language processing, and automation.

Embedded within these rapid shifts in focus were a series of assumptions about data and analytics within organizations. I am happy to say that *Becoming a Data Head* violates many of them, and it's about time. As many who work with or closely observe these trends are beginning to admit, we have headed in some unproductive directions based on these assumptions. For the rest of this foreword, then, I'll describe five interrelated assumptions and how the ideas in this book justifiably run counter to them.

Assumption 1: Analytics, big data, and AI are wholly different phenomena.

It is assumed by many onlookers that "traditional" analytics, big data, and AI are separate and different phenomena. *Becoming a Data Head*, however, correctly adopts the view that they are highly interrelated. All of them involve statistical thinking. Traditional analytics approaches like regression analysis are used in all three, as are data visualization techniques. Predictive analytics is basically the same thing as supervised machine learning. And most techniques for data analysis work on any size of dataset. In short, a good Data Head can work effectively across all three, and spending a lot of time focusing on the differences among them isn't terribly productive.

Assumption 2: Data scientists are the only people who can play in this sandbox.

We have lionized data scientists and have often made the assumption that they are the only people who can work effectively with data and analytics. However, there is a nascent but important move toward the democratization of these ideas; increasing numbers of organizations are empowering "citizen data scientists." Automated machine learning tools make it easier to create models that do an excellent job of predicting. There is still a need, of course, for professional data scientists to develop new algorithms and check the work of the citizens who do complex analysis. But organizations that democratize analytics and data science—putting their "amateur" Data Heads to work—can greatly increase their overall use of these important capabilities.

Assumption 3: Data scientists are "unicorns" who have all the skills needed for these activities.

We have assumed that data scientists—those trained in and focused upon the development and coding of models—are also able to perform all the other tasks that are required for full implementation of those models. In other words, we think they are "unicorns" who can do it all. But such unicorns don't exist at all, or exist only in small numbers. Data Heads who not only understand the rudiments of data science, but also know the business, can manage projects effectively, and are excellent at building business relationships will be extremely valuable in data science projects. They can be productive members of data science teams and increase the likelihood that data science projects will lead to business value.

Assumption 4: You need to have a really high quantitative IQ and lots of training to succeed with data and analytics.

A related assumption is that in order to do data science work, a person has to be very well trained in the field and that a Data Head requires a head that is very good with numbers. Both quantitative training and aptitude certainly help, but *Becoming a Data Head* argues—and I agree—that a motivated learner can master enough of data and analytics to be quite useful on data science projects. This is in part because the general principles of statistical analysis are by no means rocket science, and also because "being useful" on data science projects doesn't require an extremely high level of data and analytics mastery. Working with professional data scientists or automated AI programs only requires the ability and the curiosity to ask good questions, to make connections between business issues and quantitative results, and to look out for dubious assumptions.

Assumption 5: If you didn't study mostly quantitative fields in college or graduate school, it's too late for you to learn what you need to work with data and analytics.

This assumption is supported by survey data; in a 2019 survey report from Splunk of about 1300 global executives, virtually every respondent (98%) agreed that data skills are important to the jobs of tomorrow.[1] 81% of the executives agree that data skills are required to become a senior leader in their companies, and 85% agree that data skills will become more valuable in their firms. Nonetheless, 67% say they are not comfortable accessing or using data themselves, 73% feel that data skills are harder to learn than other business skills, and 53% believe they are too old to learn data skills. This "data defeatism" is damaging to individuals and organizations, and neither the authors of this book nor I believe it is warranted. Peruse the pages following this foreword, and you will see that no rocket science is involved!

So forget these false assumptions, and turn yourself into a Data Head. You'll become a more valuable employee and make your organization more successful. This is the way the world is going, so it's time to get with the program and learn more about data and analytics. I think you will find the process—and the reading of *Becoming a Data Head*—more rewarding and more pleasant than you may imagine.

Thomas H. Davenport
Distinguished Professor, Babson College
Visiting Professor, Oxford Saïd Business School
Research Fellow, MIT Initiative on the Digital Economy
Author of *Competing on Analytics, Big Data @ Work,*
and *The AI Advantage*

---

[1] Splunk Inc., "The State of Dark Data," 2019, www.splunk.com/en_us/form/the-state-of-dark-data.html.

Introduction

Data is perhaps the single most important aspect to your job, whether you want it to be or not. And you're likely reading this book because you want to be able to understand what it's all about.

To begin, it's worth stating what has almost become cliché: we create and consume more information than ever before. Without a doubt, we are in the age of data. And this age of data has created an entire industry of promises, buzzwords, and products many of which you, your managers, colleagues, and subordinates are or will be using. But, despite the claims and proliferation of data promises and products, data science projects are failing at alarming rates.[1]

To be sure, we're not saying all data promises are empty or all products are terrible. Rather, to truly get your head around this space, you must embrace a fundamental truth: this stuff is complex. Working with data is about numbers, nuance, and uncertainty. Data is important, yes, but it's rarely simple. And yet, there is an entire industry that would have us think otherwise. An industry that promises certainty in an uncertain world and plays on companies' fear of missing out. We, the authors, call this the Data Science Industrial Complex.

THE DATA SCIENCE INDUSTRIAL COMPLEX

It's a problem for everyone involved. Businesses endlessly pursue products that will do their thinking for them. Managers hire analytics professionals who really aren't. Data scientists are hired to work in companies that aren't ready for them. Executives are forced to listen to technobabble and pretend to understand. Projects stall. Money is wasted.

Meanwhile, the Data Science Industrial Complex is churning out new concepts faster than our ability to define and articulate the opportunities (and problems) they create. Blink, and you'll miss one. When your authors started

---

[1]Venture Beat. "87% of data science projects failing": venturebeat.com/2019/07/19/ why-do-87-of-data-science-projects-never-make-it-into-production

working together, *Big Data* was all the rage. As time went on, *data science* became the hot new topic. Since then, *machine learning, deep learning,* and *artificial intelligence* have become the next focus.

To the curious and critical thinkers among us, something doesn't sit well. Are the problems really new? Or are these new definitions just rebranding old problems?

The answer, of course, is yes to both.

But the bigger question we hope you're asking yourself is, *How can I think and speak critically about data?*

Let us show you how.

By reading this book, you'll learn the tools, terms, and thinking necessary to navigate the Data Science Industrial Complex. You'll understand data and its challenges at a deeper level. You'll be able to think critically about the data and results you come across, and you'll be able to speak intelligently about all things data.

In short, you'll become a *Data Head*.

WHY WE CARE

Before we get into the details, it's worth discussing why your authors, Alex and Jordan, care so much about this topic. In this section, we share two important examples of how data affected society at large and impacted us personally.

The Subprime Mortgage Crises

We were fresh out of college when the subprime mortgage crisis hit. We both landed jobs in 2009 for the Air Force, at a time when jobs were hard to find. We were both lucky. We had an in-demand skill: working with data. We had our hands in data every single day, working to operationalize research from Air Force analysts and scientists into products the government could use. Our hiring would be a harbinger of the focus the country would soon place on the types of roles we filled. As two data workers, we looked on the mortgage crisis with interest and curiosity.

The subprime mortgage crises had a lot of contributing factors behind it.[2] In our attempt to offer it up as an example here, we don't want to negate other factors. However, put simply, we see it as a major data failure. Banks and investors created models to understand the value of mortgage-backed

---

[2] www.brookings.edu/wp-content/uploads/2016/06/11_origins_crisis_baily_litan.pdf

collateralized debt obligations (CDOs). You might remember those as the investment vehicles behind the United States' market collapse.

Mortgage-backed CDOs were thought to be a safe investment because they spread the risk associated with loan default across multiple investment units. The idea was that in a portfolio of mortgages, if only a few went into default, this would not materially affect the underlying value of the entire portfolio.

And yet, upon reflection we know that some fundamental underlying assumptions were wrong. Chief among them were that default outcomes were independent events. If Person A defaults on a loan, it wouldn't impact Person B's risk of default. We would all soon learn defaults functioned more like dominoes where a previous default could predict further defaults. When one mortgage defaulted, the property values surrounding the home dropped, and the risk of defaults on those homes increased. The default effectively dragged the neighboring houses down into a sinkhole.

Assuming independence when events are in fact connected is a common error in statistics.

But let's go further into this story. Investment banks created models that overvalued these investments. A model, which we'll talk about later in the book, is a deliberate oversimplification of reality. It uses assumptions about the real world in an attempt to understand and make predictions about certain phenomena.

And who were these people who created and understood these models? They were the people who would lay the groundwork for what today we call the data scientist. Our kind of people. Statisticians, economists, physicists—folks who did machine learning, artificial intelligence, and statistics. They worked with data. And they were smart. Super smart.

And yet, something went wrong. Did they not ask the correct questions of their work? Were disclosures of risk lost in a game of telephone from the analysts to the decision makers, with uncertainty being stripped away piece by piece, giving an illusion of a perfectly predictable housing market? Did the people involved flat out lie about results?

More personal to us, how could we avoid similar mistakes in our own work?

We had many questions and could only speculate the answers, but one thing was clear—this was a large-scale data disaster at work. And it wouldn't be the last.

The 2016 United States General Election

On November 8, 2016, the Republican candidate, Donald J. Trump, won the general election of the United States beating the assumed front-runner and

Democratic challenger, Hillary Clinton. For the political pollsters this came as a shock. Their models hadn't predicted his win. And this was supposed to be the year for election prediction.

In 2008, Nate Silver's FiveThirtyEight blog—then part of *The New York Times*—had done a fantastic job predicting Barack Obama's win. At the time, pundits were skeptical that his forecasting algorithm could accurately predict the election. In 2012, once again, Nate Silver was front and center predicting another win for Barack Obama.

By this point, the business world was starting to embrace data and hire *data scientists*. The successful prediction by Nate Silver of Barack Obama's reelection only reinforced the importance and perhaps oracle-like abilities of forecasting with data. Articles in business magazines warned executives to adopt data or be swallowed by a data-driven competitor. The Data Science Industrial Complex was in full force.

By 2016, every major news outlet had invested in a prediction algorithm to forecast the general election outcome. The vast, vast majority of them by and large suggested an overwhelming victory for the Democratic candidate, Hillary Clinton. Oh, how wrong they were.

Let's contrast how wrong they were as we compare it against the subprime mortgage crisis. One could argue that we learned a lot from the past. That interest in data science would give rise to avoiding past mistakes. Yes, it's true: since 2008—and 2012—news organizations hired data scientists, invested in polling research, created data teams, and spent more money ensuring they received good data.

Which begs the question: with all that time, money, effort, and education—what happened?[3]

Our Hypothesis

Why do data problems like this occur? We assign three causes: hard problems, lack of critical thinking, and poor communication.

First (as we said earlier), this stuff is complex. Many data problems are fundamentally difficult. Even with lots of data, the right tools and techniques, and the smartest analysts, mistakes happen. Predictions can and will be wrong. This is not a criticism of data and statistics. It's simply reality.

Second, some analysts and stakeholders stopped thinking critically about data problems. The Data Science Industrial Complex, in its hubris, painted a picture of certainty and simplicity, and a subset of people drank the proverbial

---

[3] Nate Silver wrote a series of articles describing this in great detail (fivethirtyeight .com/tag/the-real-story-of-2016). Pollsters wrongly assuming independence, just like in the mortgage crisis, was one mistake.

"Kool-Aid." Perhaps it's human nature—people don't want to admit they don't know what is going to happen. But a key part of thinking about and using data correctly is recognizing wrong decisions can happen. This means communicating and understanding risks and uncertainties. Somehow this message got lost. While we'd hope the tremendous progress in research and methods in data and analysis would sharpen everyone's critical thinking, it caused some to turn it off.

The third reason we think data problems continue to occur is poor communication between data scientists and decision makers. Even with the best intentions, results are often lost in translation. Decision makers don't speak the language because no one bothered to teach data literacy. And, frankly, data workers don't always explain things well. There's a communication gap.

DATA IN THE WORKPLACE

Your data problems might not bring down the global economy or incorrectly predict the next president of the United States, but the context of these stories is important. If miscommunication, misunderstanding, and lapses in critical thinking occur when the world is watching, they're probably happening in your workplace. In most cases, these are micro failures subtly reinforcing a culture without data literacy.

We know it's happened in our workplace, and it was partly our fault.

The Boardroom Scene

Fans of science fiction and adventure movies know this scene all too well: The hero is faced with a seemingly unsurmountable task and the world's leaders and scientists are brought together to discuss the situation. One scientist, the nerdiest among the group, proposes an idea dropping esoteric jargon before the general barks, "Speak English!" At this point, the viewer receives some exposition that explains what was meant. The idea of this plot point is to translate what is otherwise mission-critical information into something not just our hero—but the viewer—can understand.

We've discussed this movie trope often in our roles as researchers for the federal government. Why? Because it never seemed to unfold this way. In fact, what we saw early in our careers was often the opposite of this movie moment.

We presented our work to blank stares, listless head nodding, and occasional heavy eyelids. We watched as confused audiences seemed to receive what we were saying without question. They were either impressed by how smart we seemed or bored because they didn't get it. No one demanded we repeat what was said in a language everyone could understand. We saw something unfold that was dramatically different. It often unfolded like this:

> Us: *"Based on our supervised learning analysis of the binary response variable using multiple logistic regression, we found an out-of-sample performance of 0.76 specificity and several statistically significant independent variables using alpha equal to 0.05."*
>
> *Business Professional: *awkward silence**
>
> Us: *"Does that make sense?"*
>
> *Business Professional: *more silence**
>
> Us: *"Any questions?"*
>
> *Business Professional: "No questions at the moment."*
>
> *Business Professional's internal monologue: "What the hell are they talking about?"*

If you watched this unfold in a movie, you might think *wait, let's rewind, perhaps I forgot something.* But in real life, where choices are truly mission critical, this rarely happens. We don't rewind. We don't ask for clarification.

In hindsight, our presentations were too technical. Part of the reason was pure stubbornness—before the mortgage crisis, as we learned, technical details were oversimplified; analysts were brought in to tell decision makers what they wanted to hear—and we were not going to play that game. Our audiences would *listen* to us.

But we overcorrected. Audiences couldn't think critically about our work because they didn't understand what we said.

We thought to ourselves there's got to be a better way. We wanted to make a difference with our work. So we started practicing explaining complex statistical concepts to each other and to other audiences. And we started researching what others thought about our explanations.

We discovered a middle ground between data workers and business professionals where honest discussions about data can take place without being too technical or too simplified. It involves both sides thinking more critically about data problems, large or small. That's what this book is about.

YOU *CAN* UNDERSTAND THE BIG PICTURE

To become better at understanding and working with data you will need to be open to learning seemingly complicated data concepts. And, even if you already know these concepts, we'll teach you how to translate them to your audience of stakeholders.

You'll also have to embrace the side of data that's not often talked about—how, in many companies, it largely fails. You'll build intuition, appreciation, and healthy skepticism of the numbers and terms you come across. It may seem like a daunting task, but this book will show you how. And you won't need to code or have a Ph.D.

With clear explanations, thought exercises, and analogies, we will help you develop a mental framework of data science, statistics, and machine learning.

Let's do just that in the following example.

Classifying Restaurants

Imagine you're on a walk and pass by an empty store front with the sign "New Restaurant: Coming Soon." You're tired of eating at national chains and are always on the lookout for new, locally owned restaurants, so you can't help but wonder, "Will this be a new local restaurant?"

Let's pose this question more formally: Do you predict the new restaurant will be a chain restaurant or an independent restaurant?

Take a guess. (Seriously, take a guess before moving on.)

If this scenario happened in real life, you'd have a pretty good hunch in a split second. If you're in a trendy neighborhood, surrounded by local pubs and eateries, you'd guess independent. If you're next to an interstate highway and near a shopping mall, you'd guess chain.

But when we asked the question, you hesitated. *They didn't give me enough information*, you thought. And you were right. We didn't give you any data to make a decision.

Lesson learned: Informed decisions require data.

Now look at the data in the first image on the next page. The new restaurant is marked with an X, the Cs indicate chain restaurants, and the Is indicate independent, local eateries. What would you guess this time?

Most people guess (I) because most of the surrounding restaurants are also (I). But notice not all restaurants in the neighborhood are independent. If we asked you to rate your confidence[4] in your prediction between 0 and 100, we'd expect it to be high but not 100. It's certainly possible another chain restaurant is coming to the neighborhood.

Lesson learned: Predictions should never be 100% confident.

---

[4]Note to our fellow statisticians: We just mean regular confidence, not statistical confidence.

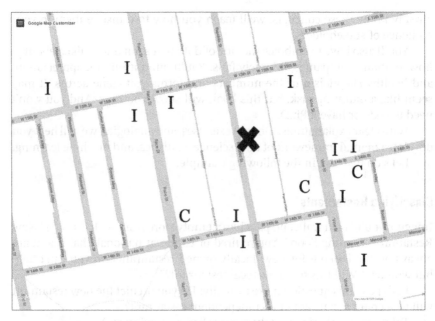

Over the Rhine neighborhood, Cincinnati, Ohio

Next, look at the data in the following image. This area includes a large shopping mall, and most restaurants in the area are chains. When asked to predict chain or independent, the majority choose (C). But we love when someone chooses (I) because it highlights several important lessons.

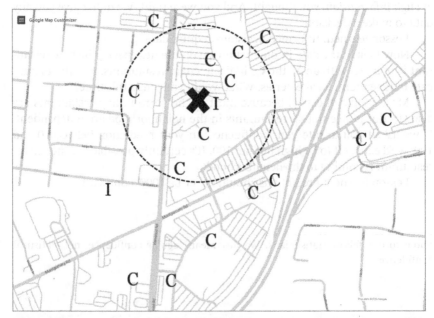

Kenwood Towne Centre, Cincinnati, Ohio

During this thought experiment, everyone creates a slightly different *algorithm* in their head. Of course, everyone looks at the markers surrounding the point of interest, X, to understand the neighborhood, but at some point, you must decide when a restaurant is too far away to influence your prediction. At one extreme (and we see it happen), someone looks at the restaurant's single closest neighbor, in this case an independent restaurant, and bases their prediction on it: "The nearest neighbor to X is an (I), so my prediction is (I)."

Most people, however, look at several neighboring restaurants. The second image shows a circle surrounding the new restaurant containing its seven nearest neighbors. You probably chose a different number, but we chose 7, and 6 out of the 7 are (C) chains, so we'd predict (C).

So What?

If you understand the restaurant example, you're well on your way to becoming a Data Head. Let's reveal what you learned, little by little:

- You performed *classification* by predicting the *label* (chain or independent) on a new restaurant by *training* an algorithm using a set of data (restaurants' location and their chain/independent label).
- This is precisely *machine learning*! You just didn't build the algorithm on a computer—you used your head.
- Specifically, this is a type of machine learning called *supervised learning*. It was "supervised" because you knew the existing restaurants were (C) chain or (I) independent. The *labels* directed (i.e., supervised) your thinking about how restaurant location is related to whether it's a chain or not.
- Even more specifically, you performed a *supervised learning classification* algorithm called *K-nearest-neighbor*.[5] If K = 1, look at the closest restaurant and that's your prediction. If K = 7, look at the 7 closest restaurants and predict the majority. It's an intuitive and powerful algorithm. And it's not magic.
- You also learned you need data to make informed decisions. Realize, however, that you need more than that. After all, this book is about critical thinking. We want to show how stuff works but also how it fails. If we asked you to predict, given the data in this Introduction's images, if the new restaurant would be kid-friendly, you wouldn't be able to answer. To make informed decisions, not just any data will do. You need accurate, relevant, and enough data.

---

[5] K-nearest-neighbor can also be used to predict numbers instead of classes. These are called *regression* problems, and we'll cover them later in the book.

- Remember the technobabble we wrote earlier? *". . . supervised learning analysis of the binary response variable . . ."*? Congratulations, you just did a supervised learning analysis of a binary response variable. Response variable is another name for *label*, and it's binary because there were two of them, (C) and (I).

You learned a lot in this section, and you did it without even realizing it.

WHO THIS BOOK IS WRITTEN FOR

As established at the beginning of this book, data touches the lives of many of today's corporate workers. We came up with the following avatars to represent people who will benefit from becoming a *Data Head*:

Michelle is a marketing professional who works side-by-side with a data analyst. She develops the marketing initiatives and her data coworker collects data and measures the initiatives' impact. Michelle thinks they can do more innovative work, but she can't articulate her data and analysis needs effectively to her data coworker. Communication between the two is a challenge. She's Googled some of the buzzwords floating around (machine learning and predictive analytics), but most of the articles she found used overly technical definitions, contained indecipherable computer code, or were advertisements for analytical software or consultation services. Her search left her feeling more anxious and confused than before.

Doug has a Ph.D. in the life sciences and works for a large corporation in its Research & Development division. Skeptical by nature, he wonders if these latest data trends are akin to snake oil. But Doug mutes his skepticism in the workplace, especially around his new director who wears a "Data is the New Bacon" t-shirt; he doesn't want to be viewed as a data luddite. At the same time, he's feeling left behind and decides to learn what all the fuss is about.

Regina is a C-level executive who is well-aware of the latest trends in data science. She oversees her company's new Data Science Division and interacts with senior data scientists on a regular basis. Regina trusts her data scientists and champions their work, but she'd like to have a deeper understanding of what they do because she's frequently presenting and defending her team's work to the company's board of directors. Regina is also tasked with vetting new technology software for the company. She suspects some of the vendors' claims about "artificial intelligence" are too good to be true and wants to arm herself with more technical knowledge to separate marketing claims from reality.

Nelson manages three data scientists in his new role. A computer scientist by training, Nelson knows how to write software and work with data, but he's new to statistics (other than one class he took in college) and machine learning. Given his somewhat related technical background, he's willing and able to learn the details, but simply can't find time. His management has also been pushing his team to "do more machine learning," but at this point, it all seems like a magic black box. Nelson is searching for material to help him build credibility within his team and recognize what problems can and cannot be solved with machine learning.

Hopefully, you can identify with one or more of these personas. The common thread among them, and likely you, is the desire to become a better "consumer" of the data and analytics you come across.

We also created an avatar to represent people who should read this book but probably won't (because every story needs a villain):

George: A mid-level manager, George reads the latest business articles about artificial intelligence and forwards his favorites up and down his management chain as evidence of his technical trendiness. But in the boardroom, he prides himself on "going with his gut." George likes his data scientists to spoon-feed him the numbers in one or two slides, max. When the analysis agrees with what he (and his gut) decided before he commissioned the study, he moves it up the chain and boasts to his peers about enabling an "Artificial Intelligence Enterprise." If the analysis disagrees with his gut feeling, he interrogates his data scientists with a series of nebulous questions and sends them on a wild goose chase until they find the "evidence" he needs to push his project forward.

Don't be like George. If you know a "George," recommend this book and say they reminded you of "Regina."

WHY WE WROTE THIS BOOK

We think a lot of people, like our avatars, want to learn about data and don't know where to start. Existing books in data science and statistics span a wide spectrum. On one side of the spectrum are non-technical books extolling the virtues and promise of data. Some of them are better than others. Even the best ones feel like the modern-day business books. But many of them are written by journalists looking to add drama around the rise of data.

These books describe how specific business problems were solved by looking at a problem through the lens of data. And they might even use words like artificial intelligence, machine learning, and the like. Don't get us wrong, these books create awareness. However, they don't delve deeply into what was done, instead focusing specifically on the problem and the solution at a high level.

On the other side of the spectrum are highly technical books. These hard-bound, 500-page tomes are as intimidating physically as the content inside is intimidating mentally.

The far sides of this spectrum have mountains of books. This perpetuates the communication gap—most people either read just the business books or just the technical books. Not both.

Thankfully, the gorge between the two extremes contains a handful of excellent books. Two of our favorites are:

- *Data Science for Business: What You Need to Know about Data Mining and Data-Analytic Thinking*, by Foster Provost and Tom Fawcett (O'Reilly Media, 2013)
- *Data Smart: Using Data Science to Transform Information into Insight*, by John W. Foreman (Wiley, 2013)

We want to add one more to this list by writing a book you can read casually without a computer or pad of paper nearby. If you enjoy our book, we highly recommend taking the next step by reading one or both of the books listed to solidify your understanding. You won't regret it.

Plus, we love this stuff. If we can convey that to you and motivate you to learn more about data and analytics—and inspire you to *want* to learn more—we'll consider this book a success.

WHAT YOU'LL LEARN

This book will help you construct a mental model of data science, statistics, and machine learning. What is a mental model? It's "a simplified representation of the most important parts of some problem domain that is good enough to enable problem solving."[6] Think of it as a new storage room in your brain where you can put information.

---

[6] This idea is discussed in an amazingly helpful book: Wilson, G. (2019). *Teaching tech together*. CRC Press.

Some books and articles start with a list of definitions: "Machine Learning is . . .", "Deep Learning is . . .", etc. Seeing a list of technical definitions without a mental model to fit the information into would be like someone dropping off boxes of clothes when you don't have a place to store them. Sooner or later, it's all going to end up in the garbage.

But with a newly constructed mental model, you will learn how to think, speak, and understand data. You'll become a *Data Head.*

Specifically, by reading this book, you will be able to:

- Think statistically and understand the role variation plays in your life and decision making.
- Become data literate—speak intelligently and ask the right questions about the statistics and results you encounter in the workplace.
- Understand what's really going on with machine learning, text analytics, deep learning, and artificial intelligence.
- Avoid common pitfalls when working with and interpreting data.

HOW THIS BOOK IS ORGANIZED

Data Heads are people who know how to think critically about data, regardless of their official role. A Data Head can be the analyst behind the keyboard doing the work, or the person at the head of the boardroom table reviewing the work of others. This book will put you, the Data Head, in various roles at different points.

While the "story" of the book is chronological, each chapter is effectively a standalone lesson and could be read out of order. But we recommend reading the book from beginning to end to help construct your mental model to go from the basics to deep learning.

The book is organized into four parts:

Part I: Thinking Like a Data Head   In this part, you'll learn to think like a Data Head—to think critically and ask the right questions about the data projects your organization takes on; what data is and the right lingo to use; and, how to view the world through a statistical lens.

Part II: Speaking Like a Data Head   Data Heads are active participants in important data conversations. This part will teach you how to "argue" with data and what questions to ask to make sense of the statistics you encounter. You'll be exposed to basic statistics and probability concepts required to understand and challenge the results you see.

Part III: Understanding the Data Scientist's Toolbox   Data Heads understand the fundamental concepts of how statistical and machine learning models work. You'll gain an intuitive understanding of unsupervised learning, regression, classification, text analytics, and deep learning.

Part IV: Ensuring Success   Data Heads understand the common mistakes and traps when working with data. You'll learn about technical pitfalls that cause projects to fail, and you'll learn about the people and personalities involved in data projects. Finally, we provide direction on how to succeed as a Data Head.

ONE LAST THING BEFORE WE BEGIN

We've established that the data field is growing faster than we can articulate the problems and opportunities it creates. We showed that our past (both society's and the authors') is filled with data failures. And only by understanding that past can we understand the future. We started you down this path by introducing you to several important concepts in the restaurant classification example.

To understand data at a deeper level, you'll need to cut through the noise, think critically about data problems, and communicate effectively with data workers. Armed with this knowledge, we know you'll be well off.

Are you ready? Your journey to become a Data Head begins on the next page.

Thinking Like a Data Head

Many companies rush to try the "next big thing" in data without ever pausing to ask the right business questions. Or learn basic data terminology. Or learn how to look at the world through a statistical lens.

Data Heads won't have that problem. Part I, "Thinking Like a Data Head," prepares you for the road ahead and puts you in the right mindset to think about and understand data. Here's what we'll cover:

I

What Is the Problem?

"A problem well stated is a problem half solved."

—Charles Kettering, inventor & engineer

The first step on your journey to become a Data Head is to help your organization work on data problems that matter.

That may sound obvious, but we suspect many of you have looked on as companies talked about how great data is but then went on to overpromise impact, misinterpret results, or invest in data technologies that didn't add business value. It often seems as if data projects are undertaken because companies like the sound of what they are implementing without fully understanding why the project itself is important.

This interaction leads to wasted time and money and can cause backlash against future data projects. Indeed, in a rush to find the hidden value in data many companies expect, they often fail at the first step in the process: defining a business problem.[1] So, in this chapter, we go back to the start.

In the next sections, we'll look at the helpful questions Data Heads should ask to make sure what you're working on matters. We'll then share an

---

[1] A robust data strategy can help companies mitigate these issues. Of course, an important component of any data strategy is to solve meaningful problems, and that's our focus in this chapter. If you'd like to learn more about high-level data strategy, see Jagare, U. (2019). *Data science strategy for dummies*. John Wiley & Sons.

example where not asking these questions leads to a project failure. Finally, we'll discuss some of the hidden costs of not clearly defining a problem right from the start.

QUESTIONS A DATA HEAD SHOULD ASK

In our experience, going back to first principles and asking the fundamental questions required to solve a problem is easier said than done. Every company has a unique culture, and team dynamics don't always lend themselves to openly asking questions, especially ones that might make others feel undermined. And many of those becoming Data Heads find that they don't have the space to even begin asking the important questions that will drive the projects forward. Which is why having a culture in which to ask these questions is as important as the questions themselves.

There's no one-size-fits-all formula for every company and every Data Head. If you are a leader, we ask that you create an open environment that will get the questions going. (This starts with inviting the technical experts into the room.) And ask questions yourself. This exhibits humility, a key leadership trait, and encourages others to join in. If you are more junior, we encourage you to try your best to ask these questions anyway, even if you're concerned it might upset the status quo. Our advice is to simply do your best. From experience, we believe simply asking the right questions always goes a lot further than not.

We want you to be prepared in the right way, trained to spot project warning signs and raise concerns at the outset. With that, we introduce five questions you should ask before attacking a data problem:

1. Why is this problem important?
2. Who does this problem affect?
3. What if we don't have the right data?
4. When is the project over?
5. What if we don't like the results?

Let's explain each in detail.

Why Is This Problem Important?

The first fundamental question is, "Why is this problem important?" It seems simple but it's one that's often overlooked. We get caught up in the hype of how we're going to solve the problem—and what we think it can do—before the project even starts. At the end of this chapter, we'll talk about the true underlying effects of not answering this question. But at a minimum, this question sets the expectations for why a project should be undertaken. This is important as data projects take time and attention—and often additional investments in technology and data. Simply identifying the importance of the problem before starting it will help optimize how company resources are best used.

You can ask the question in different ways:

- What keeps you (us) up at night?
- Why does this matter?
- Is this a new problem, or has it been solved already?
- What is the size of the prize? (What's the return on investment?)

You'll want to understand how each person sees the problem. This will help you create alignment on how everyone will end up supporting the project to solve the problem—and if they agree it should start.

During these initial discussions, you'll want to keep the focus on the central business problem and pay close attention if you hear rumblings of recent technology trends. Talk of technical trends can quickly derail the meeting away from its business focus. Be on the lookout for two warning signs:

- **Methodology focus:** In this trope, companies simply think trying some new analysis method or technology will set them apart. You've heard this marketing fluff before: "If you're not using Artificial Intelligence (AI), you're already behind . . . ." Or, companies find some other buzzword they would like to incorporate (e.g., "sentiment analysis").
- **Deliverable focus:** Some projects go off track because companies focus too much on what the deliverable will be. They say the project needs to have an interactive dashboard, for example. You start the project, but the outcome becomes about the installation of the new dashboard or business intelligence system. Project teams need to take a step back and trace how what they want to build brings value to the organization.

It may come as a surprise—or a relief—that both warnings involve technology and how it should *not* be included when defining the problem. To be clear, at some point in the project, methodologies and deliverables enter the picture. To start, however, the problem should be in direct, clear language everyone can understand. Which is why we recommend you scrap the technical terminology and marketing rhetoric. Start with the problem to be solved, not the technology to be used.

Why does this matter? We've noticed project teams have a mix of people who are enamored by data or intimidated by it. Once the problem definition conversation steers toward analysis methods or technology, two things happen. First, anyone intimidated by data might freeze up and stop contributing to the discussion—defining the business problem. Second, those enamored by data quickly splinter the problem into technical subproblems that may or may not align to an actual business objective. Once the business problem morphs into data science subproblems, it may be weeks or months before failure is discovered. No one will want to revisit the main problem once the project work starts.

Fundamentally, teams must answer "Is this a real business problem that is worth solving, or are we doing data science for its own sake?" This is a good, albeit blunt, question to ask, especially now during the hype and confusion around data science and related fields.

Who Does This Problem Affect?

The next question you'll want to ask is, "Who does this problem affect?" The spirit of the problem is not only asking who this affects, but how that person's work will be different going forward.

You should think of all layers of the organization (and perhaps its clients, if any). We don't mean the data scientist who works on the problem or the engineering team who may have to maintain software. The Data Head needs to know who the end users are. Often, it's more than just the people in the room crafting the problem, so it's super important for you to find the people whose daily work will be affected and *bring them into the meeting*.

We suggest you name names. Whose work will be different if the question gets answered? If it's many people, bring in a small group to represent them. Create a list of these people and understand how they will be affected by the project. You'll want to tie these answers back to the last question.

An exercise to help you think through this is to do a solution trial run. Assume you can answer the question, and then ask your team:

- Can we use the answer?
- Whose work will change?

This, of course, assumes you even had the right data to answer the question. (As we'll see in Chapter 4, this can be a *huge* assumption.) But you should answer these questions and go through several scenarios where the problem has been solved. In many cases, answering these questions can strengthen the project and its impact, or may identify a project with no business benefit.

What If We Don't Have the Right Data?

Every dataset has a limited amount of information inside it, and at a certain point, no technology or analysis method will help you go any further.

In the authors' experience, not asking "What if we don't have the right data?" is where companies make some of the biggest mistakes—mistakes that could be avoided if only they were considered before the project started. Because what happens is this: everyone who has worked so far on the project now wants to take it to completion no matter what. Data Heads enter the project knowing that not having the right data is a possibility. They create contingencies to pivot to collecting better data to answer the question. Or, if the data doesn't exist, they go back to the original question and attempt to redefine the project scope.

When Is the Project Over?

Many of us have been part of projects that went on too long. When expectations aren't clear before the project starts, teams wind up attending meetings out of habit and generating reports no one bothers to read. Asking "When is the project over?" before the project starts can break this trend.

The question strikes at the heart of why the project was initiated and aligns expectations. Important problems are posed because some information or product is needed in the future that does not exist today. Find out what that final deliverable is. Doing this will rekindle conversations about the project's potential return on investment and whether the team has an agreed-upon metric to measure the project's impact.

So, gather project stakeholders and identify reasons the project could end. Some reasons are obvious, like when a project ends from a lack of funding or waning interest. Set those obvious failures aside and focus on what needs to be delivered to answer the business question and conclude the project. For data projects, the final deliverable is typically an insight (e.g., "how effective was the company's last marketing campaign?") or an application (e.g., a predictive model that forecasts next week's shipping volume). Many projects will require additional work: perhaps ongoing support and maintenance, but this needs to be communicated to the team up front.

Don't assume you know the answer to this question until you've asked it.

What If We Don't Like the Results?

The last question a Data Head should ask prepares the stakeholders for something they'd rather overlook—the possibility their assumptions were wrong. "What if we don't like the results?" imagines you are at the point of no return. You've spent hours on a project only to find out the results show something different. Notice this is different from having data that *can't* answer the question. Here, the data *can* answer the question, perhaps quite confidently, but the answer is not what the stakeholders wanted.

It's never easy to get to the end of a project only to find out the results were not what you expected. This all too real scenario happens more often than we'd like to admit. Thinking first about the possibility that the project might reach an unwanted conclusion will ensure you have a plan in motion when you have to deliver the bad news.

Asking this question will also expose differences in how individuals will accept the results of the project. For instance, consider our avatar George from the introduction. George is the type of person who would ignore the results if they don't align to his beliefs, while simultaneously promoting favorable results that do. The question will hopefully uncover his bias early on before the project starts.

You don't want to start a project where you know there's only one accepted result.

UNDERSTANDING WHY DATA PROJECTS FAIL

Projects can fail for a host of reasons: lack of funding, tight timelines, wrong expertise, unreasonable expectations, and the like. Add data and analysis methods into the mix, and the list of possible failures not only grows but becomes obscured behind the analysis. A project team might apply an analysis method they can't explain on data they don't understand to solve a problem that doesn't matter—and *still* think they've succeeded.

Let's look at a scenario.

Customer Perception

You work for a Fortune 10 company, Company X, that recently received negative media attention for a socially insensitive marketing campaign. You've been assigned to a project to monitor "customer perception."

The project team consists of the following:

- The project manager (you)
- The project sponsor (the person paying for it)
- Two marketing professionals (who don't have data backgrounds)
- A young data scientist (fresh out of college and eager to apply the techniques they learned)

At the kick-off meeting, the project sponsor and data scientist quickly and excitedly discuss something called "sentiment analysis." The project sponsor heard about it at a recent tech conference after a competing company reported using it. The data scientist volunteered they knew sentiment analysis, having implemented it in their senior capstone project. They think it might be a good technique to apply to customer comments on the company's Twitter and Facebook pages. The marketers understand the technique as being able to interpret people's emotional reactions using social media data, but they don't say much.

The basic premise, you are told, is that sentiment analysis can automatically label a tweet or Facebook post as "positive" or "negative." For instance, the phrase, "Thank you for sponsoring the Olympics." is *positive*, whereas "Horrible customer service" is *negative*. Conceivably, the data scientist could count the daily totals of positives and negatives, plot the trends over time (and in real time!), and subsequently share the results via a dashboard for all to see. Most important: no one needs to read customer comments anymore. The machine will do it for everyone. So, it's decided. The project kicks off.

A month later, the data scientist proudly shows off Company X's Customer Perception Dashboard. It's updated each day to include the latest data and lists some of the week's "positive" comments along the side. Figure 1.1 zooms in on the main graphic in the dashboard: trendlines of sentiment over time. Only positive and negative values are shown, and most comments are neutral.

The project sponsor loves it. A week later, the dashboard is displayed on a monitor in the break room for all to see.

Success.

Six months later, the break room is renovated, and the monitor is removed.

No one notices.

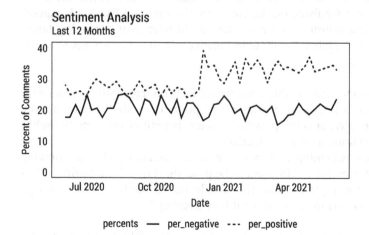

FIGURE 1.1 Sentiment analysis trends

A postmortem of the project revealed no one in the company used the analysis, not even the marketers on the team. When asked why, the marketers admit they weren't really comfortable with the original analysis. Yes, it was possible to label each communication as positive or negative. But the idea that nobody would need to read comments anymore seemed like wishful thinking. They questioned the degree to which the labeling process had even been useful. Further, they countered that perception couldn't only be measured by online interaction even if that was the dataset most readily available to support sentiment analysis.

Discussion

In this scenario, it seemed like everything went well. But the fundamental question—*why is the project important?*—doesn't appear to have been brought up. Instead, the project team moved forward attempting to answer another question: "Can we build a dashboard to monitor the sentiment of customer feedback on the company's Twitter and Facebook pages?" The answer, of course, was yes, they could. But in the end the project wasn't useful or even important to the organization.

You would think marketers would have had more to say, but they were not identified as people who would have been affected by the project. In addition, this project exhibited two early warning signs in how the team attempted to solve the problem: methodology focus (sentiment analysis) and deliverable focus (dashboard).

Moreover, the project team in the Customer Perception scenario could have taken their problem, "Can we build a dashboard to monitor the sentiment of customer feedback on the company's Twitter and Facebook pages?" and performed a solution trial run. They could have assumed a dashboard was available and updated daily with positive/negative sentiments of social media comments:

- *Can we use the answer?* The team would be thinking about the relevance of sentiment analysis on customer perception. How can the team use the information? What is the possible business benefit of knowing the sentiment of customers on social media?
- *Whose work will change?* Suppose the team convinces itself that knowing sentiment is important in order to be good stewards of the business. But is someone going to monitor this dashboard? If the trends suddenly go down, do we do anything? What if they trend up?

At this point, the marketing team would have hopefully spoken up. Would they have known what to do differently in their daily work with that kind of information? Likely not. The project, in its current form, hit a wall.

If only they asked the five questions.

WORKING ON PROBLEMS THAT MATTER

So far, we've tied project failures to not defining the underlying problem correctly. Mostly, we've placed this failure in terms of losing money, time, and energy. But there's a broader issue happening all over the data space, and it's something that you wouldn't expect.

Right now, the industry is focused on training as many data workers as possible to meet the demand. That means universities, online programs, and the like are churning out critical thinkers at lightning speed. And if working in data is all about uncovering the truth, then Data Heads want to do just that.

What does it mean, then, when they sit down to a project that doesn't whet their appetite? What does it mean for them to have to work on a poorly defined issue where their skills become bragging rights for executives but don't actually solve meaningful problems?

It means many data workers are dissatisfied at their jobs. Having them work on problems overly focused on technology with ambiguous outcomes leads to frustration and disillusionment. Kaggle.com, where data scientists from all over the world compete in data science competitions and learn new analysis methods, posted a survey and asked data scientists what barriers they face at work.[2] Several of the barriers, listed here, are directly related to poorly defined problems and improper planning:

- Lack of clear question to answer (30.4% of respondents experienced this)
- Results not used by decision makers (24.3%)
- Lack of domain expert input (19.6%)
- Expectations of project impact (15.8%)
- Integrating findings into decisions (13.6%)

This has obvious consequences. Those who aren't satisfied in their roles leave.

CHAPTER SUMMARY

The very premise and structure of this book is to teach you to ask more probing questions. It starts with the most important, and sometimes hardest, question: "What's the problem?"

In this chapter, you learned ways to refine and clarify the central business question and why problems involving data and analysis are particularly

---

[2] 2017 Kaggle Machine Learning & Data Science Survey. Data is available at www .kaggle.com/kaggle/kaggle-survey-2017. Accessed on January 12, 2021.

challenging. We shared five important questions a Data Head should ask when defining a problem. You also learned about early warning signs to spot when a question starts to go off track. If the question hints of having a (1) methodology focus or a (2) deliverable focus, it's time to hit pause.

When these questions are answered, you are ready to get to work.

CHAPTER 2

What Is Data?

"If we have data, let's look at data. If all we have are opinions, let's go with mine."

—Jim Barksdale, former Netscape CEO

Many people work with data without having a dialect for it. However, we want to ensure we're all speaking the same language to make the rest of the book easier to follow. So, in this chapter, we'll give you a brief crash course on data and data types. If you've had a basic statistics or analytics course, you'll know the terms that follow but there may be parts of our discussion not covered in your class.

DATA VS. INFORMATION

The terms *data* and *information* are often used interchangeably. In this book, however, we make a distinction between the two.

Information is derived knowledge. You can derive knowledge from many activities: measuring a process, thinking about something new, looking at art, and debating a subject. From the sensors on satellites to the neurons firing in our brains, information is continually created. Communicating and capturing that information, however, is not always simple. Some things are easily measurable while others are not. But we endeavor to communicate knowledge for the benefit of others and to store what we've learned. And one way to communicate and store information is by encoding it. When we do this, we create data. As such, *data is encoded information.*

An Example Dataset

Table 2.1 tells the story of a company. Each month, they run a different marketing campaign online, on television, or in print media (newspapers and magazines). The process they run generates new information each month. The table they've created is an encoding of this information and thus it holds *data.*

A table of data, like Table 2.1, is called a *dataset.*

Notice that it has both rows and columns that serve specific functions in how we understand the table. Each row of the table (running horizontally, under the header row) is a measured instance of associated information. In this case, it's a measured instance of information for a marketing campaign. Each column of the table (running vertically) is a list of information we're interested in, organized into a common encoding so that we can compare each instance.

The rows of each table are commonly referred to as *observations*, *records*, *tuples*, or *trials*. Columns of datasets often go by the names *features*, *fields*, *attributes*, *predictors*, or *variables.*

Know Your Audience

Data is studied in many different fields, each with their own lingo, which is why there are many names for the same things. Some data workers, when talking about the columns in a dataset, might prefer "features" while others say "variables" or "predictors." Part of being a Data Head is being able to navigate conversations within these groups and their preferences.

A *data point* is the intersection of an observation and a feature. For example, *150* units sold on 2021-02-01 is a data point.

TABLE 2.1   Example Dataset on Advertisement Spending and Revenue

Date	Ad Spending	Units Sold	Profit	Location
2021-01-01	2000	100	10452	Print
2021-02-01	1000	150	15349	Online
2021-03-01	3000	200	25095	Television
2021-04-01	1000	175	12443	Online

Table 2.1 has a header (a piece of non-numerical data) that helps us understand what each feature means. Note that not every dataset will have a header row. In such cases, the header row is implied, and the person working in the dataset must know what each feature means.

DATA TYPES

There are many ways to encode information, but data workers use a few specific types of encodings that store information and communicate results. The two most common data types are described as *numeric* or *categorical*.

Numeric data is mostly made up of numbers but might use additional symbols to identify units. Categorical data is made up of words, symbols, phrases, and (confusingly) sometimes numbers, like ZIP codes. Numeric and categorical data both split into further subcategories.

There are two main types of numeric data:

- *Continuous* data can take on any number in a number line. It represents a fundamentally uncountable set of values. Consider the weather. The outside temperature, if collected and turned into data, would represent a continuous variable. A local news station might measure a temperature of 65.62 Fahrenheit. However, they may choose to report this number to you as 65 degrees Fahrenheit, 66 degrees Fahrenheit, or 65.6 Fahrenheit.
- *Count* (or *discrete*) data, unlike continuous data, restricts the precision of the data to a whole number. For example, the number of cars you own can be 0, 1, 2, or more, but not 1.23. This reflects the underlying reality of the thing being measured.[1]

Categorical data also has two main types:

- *Ordered* (or *ordinal*) data is categorical data with an inherent order. Surveys, for example, take advantage of ordinal data when they ask you to rate your experience from 1–10. While this looks like count data, it's not possible to say the difference between survey ratings 10 and 9 is the same

---

[1]There are additional levels of continuous data, called *ratio* and *interval*. Feel free to look them up, but we rarely see the terms used in a business setting. And there are situations when the distinction between continuous and count data doesn't really matter. High count numbers, like website visits, are often considered continuous for the purpose of data analysis rather than count. It's when the count data is near zero that the distinction really matters. We'll explore this more in the coming chapters.

as the difference between 1 and 0. Of course, ordinal categorial data does not have to be encoded as numbers. Shirt size, for example, is ordinal: small, medium, large, extra-large.

■ *Unordered* (or *nominal*) categorical data does not have an underlying order to follow. Table 2.1, for example, has a Location feature with values Print, Online, Television. Other nominal variables include Yes or No responses; or Democrat or Republican party affiliation. Their order as presented is always arbitrary—it's not possible to say one category is "greater than" another.

You'll notice Table 2.1 has a Date feature, which is an additional data type that is sequential and can be used in arithmetic expressions like numeric data.

HOW DATA IS COLLECTED AND STRUCTURED

The preceding section talked about data types within a dataset, but there are larger categories to describe data that refers to how it was collected and how it's structured.

Observational vs. Experimental Data

Data can be described as observational or experimental, depending on how it's collected.

■ *Observational* data is collected based on what's seen or heard by a person or computer passively observing some process.

■ *Experimental* data is collected following the scientific method using a prescribed methodology.

Most of the data in your company, and in the world, is observational. Examples of observational data include visits to a website, sales on a given date, and the number of emails you receive each day. Sometimes it's saved for a specific purpose; other times, for no purpose at all. We've also heard the phrase "found data" to reference this type of data; it's often created as byproducts from things like sales transactions, credit card payments, Twitter posts, or Facebook likes. In that sense, it's sitting in a database somewhere, waiting to be discovered and used for something. Sometimes observational data is collected because it's free and easy to collect. But it can be deliberately collected, as with customer surveys or political polls.

Experimental data, on the other hand, is not passively collected. It's collected deliberately and methodically to answer specific questions. For these reasons, experimental data represents the gold standard of data for statisticians and researchers. To collect experimental data, you must randomly assign a *treatment* to someone or something. Clinical drug trials present a common example that generates experimental data. Patients are randomly split into two groups—a *treatment* group and a *control* group—and the treatment group is given the drug while the control group is given a placebo. The random assignment of patients should balance out information not relevant to the study (age, socioeconomic status, weight, etc.) so that two groups are as similar as possible in every way, except for the application of the treatment. This allows researchers to isolate and measure the effect of the treatment, without having to worry about potential *confounding* features that might influence the outcome of the experiment.[2]

This setup can span across industries, from drug trials to marketing campaigns. In digital marketing, web designers frequently experiment on us by designing competing layouts or advertisements on web pages. When we shop online, a coin flip happens behind the scenes to determine if you are shown one of two advertisements, call them A and B. After several thousand unknowing guinea pigs visit the site, the web designers see which had led to more "click-throughs." And because ads A and B were shown randomly, it's possible to determine which ad was better with respect to click-through rates because all other potential confounding features (time of day, type of web surfer, etc.) have been balanced out through randomization. You might hear experiments like this called "A/B tests" or "A/B experiments."

We will talk more about why this discrepancy matters in Chapter 4, "Argue with the Data."

Structured vs. Unstructured Data

Data is also said to be *structured* and *unstructured*. Structured data is like the data in your spreadsheets or in Table 2.1. It's been presented with a sense of order and structure in the form of rows and columns.

Unstructured data refers to things like text from Amazon reviews, pictures on Facebook, YouTube videos, or audio files. Unstructured data requires clever techniques to convert it into structured data required for analysis methods (see Part III of this book).

---

[2]Here's a quick example of confounding. In a drug trial, if the treatment group consists of only children and no one got sick, you'd be left wondering if their protection from the disease was caused by an effective drug treatment or because children had some inherent protection from the disease. The effect of the drug would be confounded with age. Random assignment between the control and treatment groups prevents this.

Is Data One or Many?

We should clarify where we stand on a debate you may not have known about or cared existed: Is data one or many?

The word *data* is actually the plural version of the word *datum*. (Like *criteria*—the plural of *criterion*. Or agenda—the plural of the word *agendum*.) If we were following proper rules of language, we would say "these data are continuous" and not "this data is continuous."

Both of your authors have attempted to use the correct phrasing *the data are. . .*out in the real world and it's not for us. It just sounds weird. And we're not the only ones who think so. The popular data blog FiveThirtyEight.com[3] has argued that its usage is a *mass noun*, like water or grass.

BASIC SUMMARY STATISTICS

Data does not always look like a dataset or spreadsheet. It's often in the form of summary statistics. Summary statistics enable us to understand information about a set of data.

The three most common summary statistics are mean, median, and mode, and you're probably quite familiar with them. However, we wanted to spend a few minutes discussing these statistics because we frequently see the colloquial terms "normal," "usual," "typical," or "average" used as synonyms for each of the terms. To avoid confusion, let's be clear on what each term means:

- The *mean* is the sum of all the numbers you have divided by the count of all the numbers. The effect of this operation is to give you a sense of what each observation in your series contributes to the entire sum if every observation generated the same amount. The mean is also called the *average*.
- The *median* is the midpoint of the entire data range if you sorted it in order.
- The *mode* is the most common number in the dataset.

Mean, median, and mode are called measures of location or measures of central tendency. Measures of variation—variance, range, and standard deviation—are measures of spread. The location number tells you where on the number line a typical value falls and spread tells you how spread out the other numbers are from that value.

---

[3]"Data Is" vs. "Data Are": fivethirtyeight.com/features/data-is-vs-data-are

As a trivial example, the numbers 7, 5, 4, 8, 4, 2, 9, 4, and 100 have mean 15.89, median 5, and mode 4. Notice the mean (average), 15.89, is a number that doesn't appear in the data. This happens a lot: the average number of people in a household in the United States in 2018 was 2.63; basketball star LeBron James scores an average of 27.1 points per game.

It's a common mistake for people to use the average (mean) to represent the midpoint of the data, which is the median. They assume half the numbers must be above average, and half below. This isn't true. In fact, it's common for *most* of the data to be below (or above) the average. For example, the vast majority of people have greater than the average number of fingers (likely 9.*something*).

To avoid confusion and misconceptions, we recommend sticking with mean or average, median, and mode for full transparency. Try not to use words like usual, typical, or normal.

CHAPTER SUMMARY

In this chapter, we gave you a common language to speak about your data in the workplace. Specifically, we described:

- Data, datasets, and multiple names for the rows and columns of a dataset
- Numerical data (continuous vs. count)
- Categorical data (original vs. nominal)
- Experimental vs. observational data
- Structured vs. unstructured data
- Measures of central tendency

With the correct terminologies in place, you're ready to start thinking statistically about the data you come across.

CHAPTER 3

Prepare to Think Statistically

"Statistical thinking is a different way of thinking that is part detective, skeptical, and involves alternate takes on a problem."[1]

—Frank Harrell, statistician and professor

This chapter is about how you *think* about data—about arming yourself with a mindset to consume and think critically about data you come across in your business or everyday life. It's laying important groundwork for the rest of the book, yes, but if any of these concepts are new to you, you'll likely find yourself watching the news or reading the latest popular press science articles through a new lens—*a statistical lens.*

Two important callouts before we start.

First, we're just scratching the surface here. This one chapter isn't going to replace a semester of Statistics (with apologies to any students) or dive into all the psychological aspects of "thinking" like the modern classic *Thinking, Fast and Slow.*[2] But we will introduce several concepts and establish a foundation for statistical thinking, incomplete as it may be.

Second, there is a risk in reading these next few chapters that you will become cynical about data. You might throw your arms up, claiming this statistics nonsense obfuscates the truth under complicated equations and

---

[1] F. Harrell, Professor and founding Chair of the Department of Biostatistics, Vanderbilt University: www.fharrell.com/post/introduction

[2] *Thinking, Fast and Slow,* by Daniel Kahneman (Farrar, Strauss and Giroux, 2013).

numbers, and swear off any analysis you see. Or maybe you'll throw tomatoes at every article you read because *you* learned a few tricks of the statistical trade and doubt *they* know as much as you.

Please don't. Our goal isn't to make you reject the information you see, but to question it, understand it, know its limitations, and perhaps even appreciate it.

ASK QUESTIONS

A core tenet of statistical thinking is "ask questions."

Many of us do this to some degree in our everyday life. We assume you, as a reader of a book about data, naturally question sure-bet claims from advertisers ("Lose 10 pounds in a month!" or "This hot stock is the next Amazon!") and bizarre posts on social media. (Open Facebook or Twitter, and you'll probably spot one right now.) So, the muscle is inside of you somewhere. Frankly, it can be fun to sit back as an observer and tear apart the obvious lies.

But it's much harder to do to claims and data personal to us. Every political election shows this. Take a minute to think seriously and honestly about how quickly you become suspicious of claims or numbers from the *other* political party.[3] What goes through your head? "Their sources are bad. My sources are good. Their information is false. My information is true. They just don't understand what's going on."

Clearly, this discussion can get philosophical very quickly. Our intent isn't to stir up a political debate or dive into the many factors that shape our personal and political ideologies. But there's a lesson to be learned here—it's hard to question everything when the "everything" includes your own thinking and reasoning.

More pertinent to the mission of this book, think about the information you see in the workplace. When you see data sprawled out across spreadsheets and PowerPoint presentations—information that impacts your company's success, your job performance, your possible bonus—is it viewed with any skepticism? In our experience, it's often not. Numbers in a boardroom are viewed as cold hard facts. Truth in black ink, rounded to the nearest decimal.

Why is this? Probably because you don't have time to question, poke and prod, or collect more data. This is the data you have, the data you must act on, and the data you can point to and blame if things don't work out your way. When faced with these constraints and limitations, skepticism turns off, almost reflexively. Another reason, you might suggest, is even if you

---

[3] The United States has a two-party system.

understand issues with the data, your boss probably doesn't. A chain reaction occurs where everyone assumes that someone else, up the management chain (or even down below), will take the number at face value, and that assumption permeates down to those of us staring at the spreadsheet. They'll assume it's true, so we will act on it as if it is.

Data Heads can push back. And it starts with understanding variation.

Comment on "Statistical Thinking"

We're using "statistical thinking" in a general sense, defined in the quote at the beginning of the chapter. You might prefer probabilistic thinking, statistical literacy, or mathematical thinking. No matter which phrase you prefer, all deal with the evaluation of data or evidence.

Some might wonder why this line of thinking is important. Businesses and life in general have gone on without it. So why now? Why should Data Heads care?

In an article titled, "Data Science: What the Educated Citizen Needs to Know," Harvard economist and physician Alan Garber explains why:[4]

The benefits of data science are real and have never been more salient or important. Increasingly accurate predictions will make the products of data science more valuable than ever, and will increase interest in the field. The advances can also breed complacency and blind us to flaws. Workers of the future need to recognize not only what data science does to assist them in their work, but also where and when it falls short. . . . a deeper understanding of probabilistic reasoning and the evaluation of evidence is a general skill that will serve all of them well.

THERE IS VARIATION IN ALL THINGS

Observations vary. This isn't earth-shattering news.

The stock market fluctuates daily, political poll numbers change depending on the week (and the polling firm generating the data), gas prices move up and down, and your blood pressure spikes when the doctor is present but not the nurse. Even your daily commute to work, if you broke it down and

---

[4] Link to article in the *Harvard Data Science Review*: hdsr.mitpress.mit.edu/pub/pjl0jtkp.

measured it down to the second, would be slightly different each day depending on traffic, weather, having to drop kids off at school, or stopping for a coffee. There is variation in all things. How comfortable are you with this?

You've probably accepted—or at least tolerated—this variation into your everyday life and may, in fact, be comfortable with it. (Well, maybe not the stock market swings.) Overall, though, we understand that some things change for reasons we can't always explain. And when it comes to things like filling our tires, pumping gas, or paying the electric bill, we live with these figures being different every time they're measured, so long as they make some intuitive sense. But as we established in the previous section, it's harder to put data personal to our careers or our business under the same microscope.

A business's sales fluctuate daily, weekly, monthly, and yearly. Customer satisfaction survey results can vary wildly one day to the next. If we accept the reality of variation in our lives, we don't need to explain every peak and valley. And yet, this is what businesses will attempt to do. *What was done differently the week of high sales?* leadership asks. *Let's repeat the good, reduce the bad,* they'll say. Variation makes people feel helpless about the very things they're paid to know and influence.

When it comes to business, maybe we're not as comfortable with variation as we'd like to imagine.

In fact, there are two types of variation. One type of variation stems from how data is collected or measured. This is called *measurement variation*. The second is the randomness underlying the process itself. This is called *random variation*. The difference might seem trivial at first, but this is where statistical thinking becomes important. Are decisions being made in response to random variation that can't be controlled? Or is the variation reflective of some true underlying process that, when surfaced correctly, can be controlled? We all hope for the latter.

Put simply, variation creates uncertainty.

Let's look at one hypothetical scenario and one historical case study of variation causing uncertainty.

Scenario: Customer Perception (The Sequel)[5]

You're the manager of a retailer, and the corporate office closely monitors customer satisfaction data from your store, collected from customers when they call into the 1-800-number at the bottom of a receipt. The survey asks customers to

---

[5] If it seems like we're focused on customer perception too much, it's because it is (1) hard to measure accurately, (2) highly influenced by a biased subset of people, and (3) over-scrutinized by management.

rate their satisfaction on a scale from 1–10, 10 being "completely satisfied." (A bunch of other questions are included, but it's only the first question that matters.)

To add to the ploy, corporate only wants 9s and 10s. 8s are as good as 0s. Numbers are aggregated weekly and sent to you, as the store manager, and the corporate office in a PDF file with colorful graphs whose page count is just a little too long for the amount required to deliver this information. And yet, these numbers influence your bonus, and your boss's bonus, and are nervously and obsessively pored over each week as you try to hit your 85% weekly success rate, calculated as the number of 9s and 10s divided by the total number of surveys.

We'll pause here to talk about one source of variation—how the survey measures results. Rating anything on a 1–10 scale is notoriously problematic. One person's 10 ("They didn't have what I was looking for, but an employee *helped me find a substitute!*") is another person's 5 ("They *didn't have what I was looking for!* An employee had to help me find a substitute."). We'll ignore other potential sources of variation like the presence of a rude employee, an overcrowded store, an economic downturn that has everyone on the edge, whether the customer is shopping with children, plus . . . countless others.

We're not saying that the survey itself ought to be disbanded. Rather, we want to show that the design of (that is, the way we measure) data introduces variation that we often overlook. Ignoring variation means thinking deviations from our expectations reflect low-quality service rather than expected differences inherent in the question itself. And yet businesses will attempt to chase elusive high target numbers (9s and 10s in this case) without understanding that their choices in how they measured the data was the very cause of the underlying variation.

Here's how this could play out. Suppose 50 people leave a review each day, every day, for 52 weeks. This makes 350 surveys a week and 18,200 for the year. With participation like that, you seem to have a good representation of customer perception. Then, at the end of each week, results are tallied—corporate adds up the number of 9s and 10s and divides it by the weekly total, 350—and reports the results on a graph, shown in Figure 3.1. Numbers above the 85% mark get you a pat on the back. Results below 85%, and you're sweating.

Every Monday you get the report and have a phone call with corporate about the results. Imagine the stress of these conversations in weeks 5–9. You're just below the threshold. And when you finally break above in week 10, no doubt caused by the motivation of your boss, week 11 comes along and gives you a new low. This goes on and on.

But what you're looking at in Figure 3.1 is pure randomness. We generated 18,200 random numbers that were either an 8, 9, or 10—representing how different customers vote on positive customer service—and shuffled

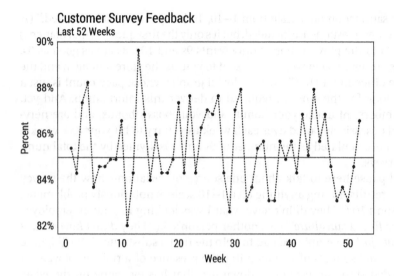

FIGURE 3.1 Weekly Customer Survey Results: Percent of Positive Reviews. The horizontal line at 85% represents the target.

them like a deck of cards.[6] Each "week" we took 350 numbers and calculated the metric. The average percentage of 9s and 10s in the dataset was 85.3% (very close to the true value of 85%), meeting the corporate standard, but each week, simply due to random variation, bounced around that threshold.

Not thinking statistically led to everyone—you, your boss, and the corporate office—chasing improvements in service to drive an arbitrary number up, even as this number was not influenced by such activities.

We term this type of activity the illusion of quantification. It's the pursuit of driving metrics without a clear statistical foundation around what they mean.

Do you see the illusion of quantification in your workplace?

Case Study: Kidney-Cancer Rates

The highest kidney-cancer rates in the United States, measured by the number of cases per 100,000 of population, occur in very rural counties sprawled out across the Midwest, South, and West regions of the country.

Pause to think about *why* this might be.

---

[6] In our simulation, the chance of an 8 rating was 15%, the chance of a 9 was 40%, and the chance of a 10 was 45%. Therefore, since we generated this data, we know the true value of the satisfaction metric, receiving a 9 or 10, is exactly 85%.

Perhaps you suspect, being in rural areas in the interior of the country, the residents have less access to adequate healthcare. Or maybe it's the result of unhealthy living caused by meat-heavy, salt-laden high-fat diets—or too much beer and spirits. It's easy, natural really, to start building narratives around facts. You can already picture researchers starting to devise remediation measures to alleviate the problem.

But here's another fact: the *lowest* kidney-cancer rates in the United States also occur in very rural counties sprawled out across the Midwest, South, and West—often neighboring the borders of the highest-rate counties.[7]

How can both be true? How can two cities with similar demographics have so very different results? Every reason you might think of to explain why very rural counties have high rates of kidney cancer would surely apply (to some degree) to their neighboring counties. So, something else must be going on.

Let's take two neighboring counties in rural Midwest, County A and County B, and assume they each have only 1000 residents. If County A has zero cases, its rate would be 0, obviously in the lowest rate category. But if County B has a single case of kidney cancer, its rate would be 100 cases per 100,000 population, giving it the highest rate in the country. It's the low population in the counties causing the high variation, simultaneously producing the highest and lowest rates. In contrast, one additional case in New York County (Manhattan, New York City), with a population over 1.5 million residents, would barely move the needle. Going from 75 cases to 76 cases would change the number of cases per 100,000 from 5 to 5.07.

In fact, this variation was real and measured in an *American Scientist* article titled, "The Most Dangerous Equation."[8] Figure 3.2 summarizes the results for U.S. counties. The sparsely populated counties, on the left side of the plot, show much higher variation in cancer rates, from 0 up to 20, the highest in the country. As population increases, moving left to right in the plot, the variation starts to reduce, giving a triangular shape. There's much less variation on the right side of the figure, indicating densely populated counties are more robust to additional cases and stabilize around 5 cases per 100,000 of population.

The article shared other examples where small numbers cause high variation. For instance, would you be surprised to learn that small schools have both the best and the worst test scores? One or two students not passing a test can cause a huge swing in overall percentages. Small numbers can lead to extreme results.

---

[7] Suppose we flip-flopped the story and told you the rural areas had the lowest kidney-cancer rates to start the case study. What reasons would you have listed? Give it a shot. You'll see just how easy it is to craft a story around data.

[8] Wainer, H. (2007). The most dangerous equation. *American Scientist*, 95(3), 249.

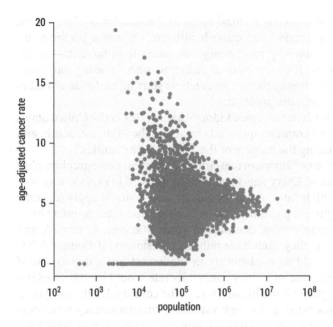

FIGURE 3.2   Reprint of *American Scientist* figure

PROBABILITIES AND STATISTICS

In the last few sections, we explained variation and talked about how it's a source of uncertainty for many businesses. Uncertainty, in fact, can be managed and this is where probability and statistics enter the picture.

We often use the terms probability and statistics interchangeably, if not together, when describing the mathematics of outcomes. But here we can go a little deeper to truly understand the difference.

Imagine a big bag of marbles. Inside, you don't know what color they are. You don't know their shape or size. You really don't even know how many marbles are in the bag, but you reach in the bag and blindly grab a handful.

Let's stop for a moment. You have a bag of marbles you haven't peeked into and a fistful of glass rolling between your fingers you haven't looked at. You really have no information about what's in your hand or in the bag.

Now here's the difference. In probability, you find out exactly what's in the bag, and use the information to guess what's in your hand. In statistics, you open your hand and use the information to tell us what's in the bag.

Probabilities drill-down; statistics drill-up. Makes sense?

Let's look at two real-life examples:

■ **Las Vegas casinos are built on probability.** Every time you play a casino game you are pulling from their bag of marbles, made up of wins and losses. There are just enough winning marbles within the casino bag to keep you interested in playing. Casinos understand variation—indeed they've commercialized it through payouts and losses optimized to keep you interested and exhilarated. Over the long term, however, casinos know they will make money because they created the bag from which all marbles are pulled, and they know exactly what's inside. With every bet made, chip laid at the table, and lever pulled on the side of a slot machine, casinos know the underlying probability of your success. If you think about how much data casinos have, you can see that they both live in a world of variation but also have a clear sense of probable outcomes.

■ **Political polling is based on statistics.** In a casino, the bag of marbles is meticulously designed and is constantly sampled from. In an election, however, politicians don't know what's really inside the entire bag until election day when all the marbles (i.e., votes) are revealed.[9] Politicians only get this one chance to learn what's in the bag—and whether it contains enough winning marbles for them. Before the election, politicians and political parties only have access to a small set of random marbles (called surveys), and they pay a lot of money for that access. Using that sample, they infer the patterns inside the bag and adjust their campaigns accordingly. Because their information is incomplete (and because they often introduce bias and error), they don't always get it right. But when they do, it's the difference between winning an election and not.

Let's take a quick look at some important concepts in probability and statistics in the following sections.

Probability vs. Intuition

Earlier in this chapter, we said that random variation cannot be controlled. But it can be measured, and probability gives us the tools to do it.

Sometimes probabilities make complete sense to us. If you've rolled a fair die or spun a dreidel, you recognize you have a known chance of landing on a specific number (1 in 6) or letter (1 in 4). Simple games of chance make a lot

---

[9]We're oversimplifying. In an election, political parties are trying to influence the makeup of the bag, both in number of marbles and their color. But even in that case, they still don't know everything inside and rely on sampling.

TABLE 3.1   Probability Dentists Agree to an Advertising Claim

	Dentists				
	1	2	3	4	5
Agreement?	Yes	Yes	Yes	Yes	No
Probability	0.8	0.8	0.8	0.8	0.2

of sense to us. Simple probabilities feel intuitive. Indeed, they make so much sense to us, that they often obscure underlying complexity. Commercials, for instance, play on appeal to simple probabilities by reducing them to something that feels like we intuitively understand it.

You've likely seen this commercial before: "4 out of 5 dentists agree" to an advertising claim, X (X can be whatever you want—chewing gum reduces cavities, or baking soda whitens teeth—it doesn't matter).

Now, suppose five dentists are sitting in front of you. Knowing that 80% of all dentists agree that X, how likely is it that exactly four out of the five dentists in front of you agree?[10]

100%? 90%? or 80%?

The actual answer is 41%.

This seems too low intuitively, but it is correct. Let's look at why. Table 3.1 shows one way a sample of five dentists could agree to X.

Probability of this combination $= 0.8 * 0.8 * 0.8 * 0.8 * 0.2 = 0.08192$

Or, for brevity,

$$p = 0.8^4 + 0.2 = 0.08192$$

But there are five different combinations of agreement, where each dentist could the one "No," shown in Table 3.2

Thus, multiply the original probability by five: $0.08192 * 5 = .4096$, or 41% for short.

Four out of five dentists may agree, on average, but that's no guarantee that in *every* sample of five dentists that four will agree to claim X. Going back to our marble analogy, if the bag of marbles contained 80% marbles with yeses and 20% noes, some handfuls of five will have all five yeses. And in the rare case, all five noes. (That's variation for you.)

We share this exercise to highlight, once again, how people underestimate variation, especially when dealing with small numbers. What people

---

[10] Example from www.johndcook.com/blog/2008/01/25/example-of-the-law-of-small-numbers

TABLE 3.2   Possible Combinations of 4 out of 5 Dentists Agreeing

	Dentists: Do you agree?				
Combination	1	2	3	4	5
1	Yes	Yes	Yes	Yes	**No**
2	Yes	Yes	Yes	**No**	Yes
3	Yes	Yes	**No**	Yes	Yes
4	Yes	**No**	Yes	Yes	Yes
5	**No**	Yes	Yes	Yes	Yes

expect to see, based on intuition, rarely matches reality when we calculate the probabilities. And *underestimating variation* causes people to *overestimate their confidence* in small data. This has been coined the "law of small numbers." It's "the lingering belief . . . small samples are highly representative of the populations from which they are drawn."[11]

Thinking statistically, like a Data Head should, means being mindful of our intuition, realizing it can play tricks on us. (We'll explore several more of these examples and misconceptions in the coming chapters.)

Discovery with Statistics

Statistics is often broken out into *descriptive* statistics and *inferential* statistics. You are probably familiar with descriptive statistics even if you don't use the phrase. Descriptive statistics are the numbers that summarize data—the numbers you read in the newspaper or see on the projection screen at work. Average sales last quarter, year-over-year increases, unemployment rates, etc. Measures like mean, median, range, variance, and standard deviation are descriptive statistics and require specific formulas to calculate. Your old Statistics book is full of them.

Descriptive statistics are a deliberate oversimplification of data—a way to condense an entire spreadsheet of company sales data into a few key measures that summarize the main information. Going back to the marble analogy, descriptive statistics is simply counting and summarizing the marbles in your hand.

While useful, we're rarely content to stop here. We want to go the extra step and understand how we can take the information in our hand and make a principled guess to infer the general contents of the entire bag. This is inferential statistics. It's the process of "going from the world to the data,

---

[11] Tversky, A., & Kahneman, D. (1974). Judgment under uncertainty: Heuristics and biases. *Science, 185*(4157), 1124-1131.

and then from the data back to the world."[12] (We will go deeper into this in Chapter 7.)

For now, let's consider an example. Imagine how you'd react seeing the headline, "75% of Americans Believe UFOs Exist!" after learning it was sampled from 20 tourists at the International UFO Museum and Research Center in Roswell, New Mexico. Do you think you can accurately *infer* the true underlying percentage of Americans who believe in UFOs based off what you now know about the study?

A Data Head's skepticism meter would go off immediately. The statistic, 75%, is not trustworthy based on the

- *Biased* sample. People visiting Roswell would be much more likely to believe in UFOs than the general public.
- Small *sample size.* You've learned how much variation small sample sizes introduce. Inferring what millions think based on 20 people doesn't make much sense.
- *Underlying assumptions.* The headline specified "Americans" as believing in UFOs simply because the test was taken in America. But the museum, as you might recall, is an international attraction. You don't know that everyone who participated in the survey is an American.

Concepts like *bias* and *sample size* are tools of statistical inference that help us understand if the statistics we see or calculate are nonsense. And they're an important part of your toolkit. The underlying assumptions are equally as important to consider, as well. Thinking like a Data Head requires that you don't take assumptions said in a conclusion at face value.

So, as you see data in your work, don't blindly acquiesce to the information you see or even the intuition you feel.

Think statistically. Ask questions. That's what Data Heads do. The upcoming chapters will tell you the questions to ask to help you think statistically.

Statistical Thinking Resources

Earlier in this chapter, we made clear that we can only scratch the surface on statistical thinking. Fortunately, several great books go into more depth about the topic. Our favorites are:

- *Damned Lies and Statistics: Untangling Numbers from the Media, Politicians, and Activists*, by Joel Best (University of California Press, 2001)

---

[12] O'Neil, C., & Schutt, R. (2013). *Doing data science: Straight talk from the frontline.* O'Reilly Media, Inc.

- *How Not to Be Wrong: The Power of Mathematical Thinking*, by Jordan Ellenberg (Penguin Books, 2015)
- *How to Lie with Statistics*, by Darrell Huff (W. W. Norton & Company, 1993)
- *Naked Statistics: Stripping the Dread from the Data*, by Charles Wheelan (W. W. Norton & Company, 2013)
- *Proofiness: How You're Being Fooled by the Numbers*, by Charles Seife (Penguin Books, 1994)
- *The Drunkard's Walk: How Randomness Rules Our Lives*, by Leonard Mlodinow (Pantheon, 2008)
- *The Signal and the Noise: Why So Many Predictions Fail—But Some Don't*, by Nate Silver (Penguin Press, 2012)
- *Thinking, Fast and Slow*, by Daniel Kahneman (Farrar, Strauss and Giroux, 2013)

CHAPTER SUMMARY

In this chapter, we laid a foundation of statistical thinking from which we'll continue to build upon throughout the book.

Specifically, we described the importance of variation and understanding how it exists within the context of the things we measure. We showed that surveys can have wide variation when they solicit customer opinion. Not because the service was bad (though, it may have been) but because the question itself predisposes wildly different responses that might be characterized as similar until measured.

We also talked about probability and statistics. They help us manage variation by demonstrating that some of it is predictable and that some of it won't even matter over the long term.

Probability drills-down: it uses a large universe of information to tell us what we'll find if we grab random scoops from it. Statistics drills-up: it tells us about the larger universe of information by using the small bits we have access to. Both probability and statistics are tools to help us learn more about when a complete picture of what we want to know remains obscure. Finally, we discussed how you can use your knowledge of probability and statistics to hone your skepticism.

Speaking Like a Data Head

Part II, "Speaking Like a Data Head," continues the previous part's charge to think statistically and question everything. In fact, Part II gives you questions to ask and things to think about whether you're viewing someone else's data project or doing the work yourself. Many of the forthcoming section headers will be named for the very same questions you'll need to ask. Consider it your reference guide for asking tough questions. Here's what we'll cover:

Chapter 4: *Argue with the Data*

Chapter 5: *Explore the Data*

Chapter 6: *Examine the Probabilities*

Chapter 7: *Challenge the Statistics*

Equipped with these chapters, you'll be able to ask intelligent questions about the data and analytics you encounter in the workplace.

Argue with the Data

"The combination of some data and an aching desire for an answer does not ensure that a reasonable answer can be extracted from a given body of data."

—John Tukey, famous statistician

As you become a Data Head, your job is to demonstrate leadership in asking questions about the data used in a project.

We're talking about the underlying raw data—the raw material—from which all statistics are calculated, machine learning models built, or dashboard visualizations created. This is the data stored in your spreadsheets or databases. If the raw data is bad, no amount of data cleaning wizardry, statistical methodology, or machine learning can hide the stench. Therefore, we can best summarize this chapter with a phrase you may have heard before, "garbage in, garbage out." In this chapter, we lay out the types of questions you should ask to find out if your data stinks.

We have identified three main prompts or questions to ask to help you argue with the data. Within those questions we offer additional follow-up questions.

- Tell me the data origin story.
  - Who collected the data?
  - How was the data collected?

- Is the data representative?
  - Is there sampling bias?
  - What did you do with outliers?
- What data am I not seeing?
  - How did you deal with missing values?
  - Can the data measure what you want it to measure?

In the sections that follow, we'll present each question, why you should ask it, and what issues it often uncovers.

Before we do that, however, let's begin with a thought exercise.

WHAT WOULD YOU DO?

You are in charge of a high-profile project for a tech company on the verge of a breakthrough in the self-driving car industry. It's an important moment for you and your work, not to mention your career. The success of your product demonstration would mean the fulfillment of many late nights, overly optimistic promises made to executives, forgiveness of project delays, and making good on those big budget research and development dollars you begged for.

And it's the night before the reveal of a new prototype automobile.

Senior executives, dozens of employees, potential investors, and media have traveled hundreds of miles to see what could be a pivotal moment in automobile history. But late in the evening, your senior engineer reports that tomorrow's forecast is a freezing 31^0 Fahrenheit. The engineer tells you that cold temperatures could compromise vital components of the car's innovative prototype self-driving system. It's not that they know for sure there could be an issue. Rather, the system—which will eventually be adapted to and stress-tested in freezing temperatures—simply hasn't been tried in the cold yet, and the demonstration is at risk for a very public and costly disaster.

But the risks of postponing are also costly. An event like this, if it doesn't happen tomorrow, is not easy to reschedule. It could be months before conditions are once again perfect. Your company has spent the better part of the last year creating excitement for this very moment. If it doesn't happen tomorrow, the excitement might never be this good.

You ask to see the data behind the engineer's concern—that temperature might compromise the car's internal components—and you're presented with the data shown in Figure 4.1.

Your engineer explains that the company has performed 23 test drives at various temperatures, and seven of them (shown in the figure) had incidents in which a critical part of the self-navigation system became distressed. Two of the test drives had two critical part failures.

Part Distress Data

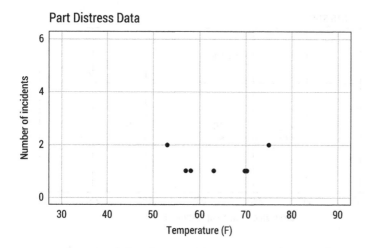

FIGURE 4.1 Plot of test drives with critical component failures as a function of temperature

Indeed, your engineers have considered the chance of component failure. That's why they created a redundancy. Each system has six critical parts (which is why the vertical y-axis on the chart goes to 6). Having backup parts means several can break before the entire car stops in failure. In 23 test runs, never has more than two components failed—and never has this caused an issue to the car's usability. Both of these cases, which happened at 53°F and 75°F, did not stop the car from running. The lowest temperature at which the test was conducted was 53°F; the highest was 81°F.

"Still, we just haven't tested the system at colder temperatures," say the engineers on your team. You hear their concerns.

As much as you try to see it, temperature doesn't seem to influence part failure except that they happened at temperatures well above 30 degrees. It's hard to picture a scenario where cold temperatures would impact more than two out of the six parts given the data being shown in the context of the 23 trial runs. And the car can run on four critical parts. If there is a failure tomorrow of two parts at most, would the world even know?

What would you do? Postpone?—or, proceed as scheduled?

Take a moment. Are there any missing data points you might want to consider?

Missing Data Disaster

On January 28th, 1986, with the world watching, NASA launched the space shuttle *Challenger* from the Kennedy Space Center in Florida in freezing temperatures.

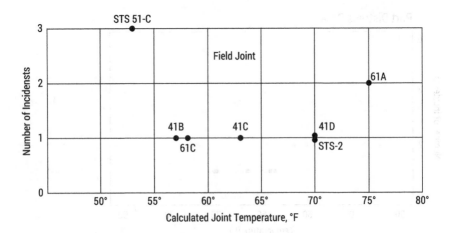

FIGURE 4.2 Plots of flights with incidents of O-ring thermal distress as a function of temperature. This figure is from the *Report of the Presidential Commission on the Space Shuttle Challenger Accident.*

Many of us know this part of the *Challenger* story. But we might not know the data story that was behind it. In fact, the *Challenger* had also had six critical parts known as O-rings, which "prevent burning rocket fuel from leaking out of booster joints."[1] There had been seven incidents of distressed O-rings in 23 trial runs leading up to the launch date.

Does this scenario sound familiar?

NASA faced the same quandary the night before as you did in your thought exercise. According to the Rogers Commission report (which was commissioned by President Ronald Reagan after the *Challenger* accident), a meeting took place the night before launch to discuss the issue.

> The managers compared as a function of temperature the flights for which thermal distress of O-rings had been observed—not the frequency of occurrence based on all flights [Figure 4.2].[2]

"In such a comparison," the report noted, "there is nothing irregular in the distribution of O-ring 'distress' over the spectrum of joint temperatures at launch between 53 degrees Fahrenheit and 75 degrees Fahrenheit."

---

[1] Quote from NRP article. "Challenger engineer who warned of shuttle disaster dies." www.npr.org/sections/thetwo-way/2016/03/21/470870426/challenger-engineer-who-warned-of-shuttle-disaster-dies

[2] Quote from *Report to the President by the Presidential Commission on the Space Shuttle Challenger Accident.* Page 146. Available online at spaceflight.nasa.gov/outreach/SignificantIncidents/assets/rogers_commission_report.pdf.

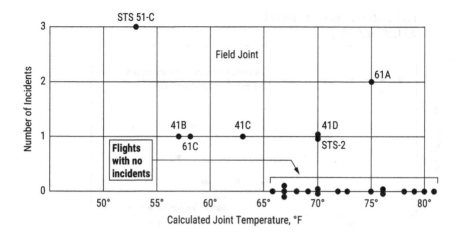

FIGURE 4.3 Plots of flights with incidents of O-ring thermal distress as a function of temperature including trial runs with no incidents. This figure is from the *Report of the Presidential Commission on the Space Shuttle Challenger Accident.*

NASA went ahead with the launch, based on their review of these failures. But on the day of the launch, the O-rings failed to seal properly in the unusually cold conditions, and the shuttle broke apart 73 seconds into its flight, killing all seven astronauts aboard.

Can you think of any data they missed?

What about the 16 test runs with no failures? Figure 4.3 shows the additional test runs, documented in the Rogers Commission.

Going back to the thought exercise—would you have thought to ask for the missing data? If you had—and perhaps gave it to statisticians to review—you might have seen there was an underlying trend predictive of part failure at lower temperatures. Figure 4.4 shows the trial runs of our autonomous automobile example, including those did not result in critical failures.

NOTE

In Chapter 2, "What Is Data?" we mentioned how the type of data dictates the analysis method. This is one of those cases. The number of incidents is numeric count data and requires a special type of modeling called *binomial regression*. Because this is count data and not continuous data, you cannot use linear regression, which you'll learn in Chapter 9. Describing binomial regression is beyond the scope of the book, but the data types dictate the analysis method. If you used linear regression to draw a straight line through the data, you'd predict negative failures for hot temperatures, which makes no sense.

Part Distress Data with Model
All Data Included

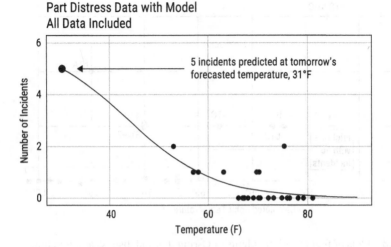

FIGURE 4.4   Plot of test drives with and without critical component failures as a function of temperature. The fitted line is a binomial regression model.

Statisticians, engineers, and researchers have studied the *Challenger* data[3] in the decades that followed. And we wanted to present to you a real-world scenario to show that these are the questions data workers must face. One article, published in the *Journal of the American Statistical Association (JASA)*, a leading statistics journal, originally presented the analysis we re-created in Figure 4.4 predicting five out of the six primary O-rings could fail in freezing temperatures. This chart used the data not originally considered the night before launch. The article makes the case that "statistical science could have provided valuable input to the launch decision process."[4]

Would you have wanted to see this same chart the night before?

---

[3] The data is available for download from the University of California, Irvine, Machine Learning Repository: archive.ics.uci.edu/ml/datasets/Challenger+USA+Space+Sh uttle+O-Ring

[4] Dalal, S. R., Fowlkes, E. B., & Hoadley, B. (1989). Risk analysis of the space shuttle: pre-Challenger prediction of failure. *Journal of the American Statistical Association, 84*(408), 945-957.

Alex's Comment on the Challenger Data

Astute readers may have noticed a slight discrepancy between the data we presented in Figure 4.1 and the plots of the data from the Rogers report in Figures 4.2 and 4.3. You'll notice the flight at 53°F in Figure 4.1 had two incidents, but Figures 4.2 and 4.3 have three. (All other data points match.) There are six primary O-rings per shuttle, and six secondary O-rings. The third incident at 53°F in Figures 4.2 and 4.3 occurred on a secondary O-ring and was the only incident of secondary O-ring damage in the 23 pre-accident flights. The analysis here focused on the six primary O-rings to match the analysis done in the *JASA* article.

The *Challenger* story offers a chilling scenario of a common phenomenon: that we often look at data that appears to encode the information we need while discarding data we assume wouldn't be relevant. Indeed, we must admit that few situations will ever be as dire as the *Challenger* scenario. Because so much was at stake, the *Challenger* scenario serves as an important story that has an immediate and obvious impact.

We don't want to speculate that if only they looked at the full dataset, they would have made the right decision. There is really no way to know this. Other factors were surely at play. Rather, we just want to point out that there are often stories to uncover when we further argue with the data.

And to that extent, the story the *Challenger* tells is clear. Most businesses don't argue with their data. Instead, they have a culture of acceptance. The effect of this is a slow burn where data projects continue to fail without important questions being asked during the project.

With that, we return to this chapter's goal of teaching you how to argue with the data and what questions to ask.

TELL ME THE DATA ORIGIN STORY

All data starts somewhere. We shouldn't take for granted its origins. So, we suggest you ask, "What is the origin of this data?"

We like this question because it's an open-ended and quick way for you to judge if the underlying raw data aligns to the question being asked of it, and it doesn't require mathematical or statistical knowledge to answer. More important, we believe the question itself creates a sense of openness and builds trust (or creates doubt) to the results that follow.

Listen carefully to the answer for potential issues of correctness and integrity stemming from the person or organization who created the data.

In particular, you are probing for answers to the following questions:

- Who collected the data?
- How was the data collected? Is it observational data or experimental?

Who Collected the Data?

When asking who collected the data, we're looking first to establish exactly where the data originated and second, if there are any issues surrounding its origin that would make us ask more follow-up questions.

Many large companies take for granted their data was collected by an internal resource. For instance, a company using workforce data—that is, data based on surveys and associated information of its own employees— might actually be using data owned and collected by a third-party vendor. That last mile of consumption of this data might happen through a portal owned by the company, giving the appearance that the data was collected and is owned by the company, even when it isn't.

In particular, we want you to identify exactly who collected the data. As a Data Head, you must question if external data is reliable and relates to the business problem at hand. Most third-party data is not readily usable in the format it's given to businesses. You, or someone on the data team, will be responsible for transforming the data from a third party into the right struc- ture and format to align with the unique data assets in your company.

How Was the Data Collected?

You should also probe for *how* data was collected. This question will help uncover if conclusions are being made about the data that aren't allowed. It will also present to you if there are underlying ethical issues with the data collection.

Recall, there are two basic data collection methods: observational and experimental.

Observational data is collected passively. Think website hits, class attend- ance, and sales numbers. Experimental data is collected under experimental conditions with treatment groups and time-tested precautions to maintain integrity and avoid confounding. Experimental data is the gold standard. Because of the care the experiment provides to ensuring the results are reli- able, this data presents an opportunity upon which to derive some causal understanding. For instance, experimental data can help answer the following questions:[5]

- If we give the patient a new drug, will it cure the disease?
- If we discount our product 15%, will it boost next quarter's sales?

Most business data, however, is observational. Observational data should not be used (at least, not exclusively) to derive causal relationships.[6] Because the data was not collected with specific care toward an experimental design, the usefulness of the data and its underlying results must be presented within this context. Any claims of causality with observational data should be met with skepticism.

Asking how the data was collected will help you uncover whether causality has been assigned when it was not possible to do so. In fact, the incorrect assignment of causality is enough of a problem that we'll return to it several times throughout this book.

It sounds simple enough to use experimental data whenever you can, but to add to the never-ending complexities of working with data, it's not always possible, cost-effective, or even ethical to collect experimental data. For example, if you were assigned to study the impact of "vaping" (smoking electronic cigarettes) on teenagers, you can't randomly assign teens to a treatment and control group and force the treatment group to vape in the name of science. That's not exactly ethical.

As a Data Head, you must work with the data you have while also mediating its ability to drive business decisions. Some companies and departments have the resources to follow up promising observational data with solid experiments. And yet, other business problems do not easily lend themselves to experiments.

IS THE DATA REPRESENTATIVE?

You need to make sure the data you have is representative of the universe you care about. If you care about shopping habits of teenagers in the United

---

[5] Note that these are the types of business questions you should be asking before you start a data science project, as outlined in Chapter 1.

[6] There are clever ways to use observational data to suggest some causal relationships. It relies on strong assumptions and clever statistics. There is a field dedicated to this study called Causal Inference.

States, the dataset you have ought to be representative of the larger universe of all teenagers' shopping habits in the country.

Inferential statistics exists precisely because we rarely, if ever, have all the data we need to solve a problem. We're forced to rely on samples.[7] But if the sample fails to be representative of what you care about, any insight you gain from the sample will not reflect the reality of the larger universe.

Here are targeted questions to ask to see if your data is representative:

- Is there sampling bias?
- What did you do with outliers?

Is There Sampling Bias?

Sampling bias happens when the data you have is consistently off or different from the data you care about. Sampling bias is often uncovered indirectly after many decisions have been made on data that is poorly representative of the problem it exists to support. It's only after those decisions continually fail to achieve what is predicted by the data do analysts go back and review if the data was the right data in the first place.

If you want to find out a politician's approval rating and only poll from voters in their political party, you have introduced sampling bias into your data. Good experimental design manages the potential for sampling bias.

In your own work, you may be faced with inherently biased data. Observational data, in particular, is susceptible to bias. The "why was this data collected?" question should uncover why the data exists for your use. Rarely has thought been put into this data to ensure it is free from bias.

You should treat all observational data as inherently biased. You don't need to throw this data out, but you must always present it within the context of its shortcomings.

What Did You Do with Outliers?

Imagine looking at a company's salary data and seeing the number $50,000,000 USD for a new hire in a management role. Would you describe this as an outlier? What would you do with it?

Outliers are defined as data points that differ significantly from other data points. The discovery of outliers should spark a discussion about which data should be rationally excluded from the analysis. Not liking what an extreme

---

[7] If you can collect every observation in a population or universe you care about, that's called a *census*.

value does to an analysis does not automatically mean it should be deleted. For a data point to be removed, have good business justification for removing it.

Arbitrarily picking and choosing which data points are outliers can introduce sampling bias. If outliers are dropped, the original data point and reason for dropping it should be documented and communicated, especially if the results changed substantially.

WHAT DATA AM I NOT SEEING?

Data that's missing either hasn't been recorded (it has no origin) or you just haven't looked at it yet. Consider the following examples:

- Data representing the underemployed is not considered in the unemployment rate.
- A mutual fund company "retires" poor performing mutual funds, making the long-term return of the remaining funds appear greater on average.
- 16 out of the 23 data points from shuttle flights were missing in the *Challenger* story.

It's worth thinking about information that hasn't been encoded in the data you can see. Play the detective.[8]

How Did You Deal with Missing Values?

Missing values are literal holes in a dataset. They represent data points that weren't collected, or may be outliers that were removed (see previous section). Although missing values are a challenge, there are methods to address them. So, it's always worth asking "How did you deal with missing values?"

Suppose you work for a credit card company and you collect data from credit card applicants: name, address, age, employment status, income, monthly housing costs, and number of bank accounts owned. Your job is to predict if these applicants will be late on payments in the next year. Several applicants, however, do not enter their income. So, the system stores it as a blank—a missing value.

Let's go back to the data's origin story. The story starts with applicants applying for a credit card. It's possible the applicant didn't provide their income because they thought they would be denied a credit card if their income was too low. This means the very presence of a missing value might be

---

[8] We'll come back to this idea in a later chapter when we discuss "survivorship bias."

predictive to whether the application may have a late payment in the future. You don't want to throw out this information!

With this additional insight, a data scientist could create a new categorical feature call "Income Present?" and enter the value 1 if the person entered their income and a 0 if they did not. In this way, you've reencoded missing data into its own categorical variable.

Can the Data Measure What You Want It to Measure?

We often believe we can measure anything and everything. But you should evaluate whether the data provided can truly measure complicated ideas. Consider, for example, the following:

- How would you measure your client's loyalty to your business?
- What data would you use to measure "brand equity" or "reputation?"
- What data can show how much you love your child? Or your pet?

These are hard things to measure. Data, by way of encoding information, enables us to get closer to these answers, but by and large the data we use is truly proxy—a stand in, as it were—for the thing we're trying to measure. The degree to which the proxy reflects reality varies.[9]

To the extent your data indirectly measures something, you should be truthful and honest about it. Measuring complicated concepts like brand equity and reputation require indirect approximations, as these things are truly hard to measure.

ARGUE WITH DATA OF ALL SIZES

It's easy to think getting more data would overcome inherent issues in working with sample data. The larger the sample, the more reliable—this is a misunderstanding in statistical thinking. If the data is collected properly, a larger sample will help, but if the data has bias, additional data can't save you.

And thus the short-lived hype around Big Data suggested that more data could itself create more scientific rigor by virtue of volume. Don't think a dataset is too big to argue with. Statistics doesn't have an absolute data-size

---

[9] Manufacturing, engineering, and research organizations should also consider studies to determine the repeatability and reproducibility of data measured by technical equipment.

threshold, that when passed, a sample no longer contains bias. Statistics deals in the tradeoffs between what you ultimately want to know with the data you have.[10]

CHAPTER SUMMARY

We opened the chapter with details of the *Challenger* accident but put you in the driver's seat. As we pointed out at the start of this book, data mistakes are made by smart people. People and organizations can and do make mistakes.

That's why we presented you with questions you should ask and the various issues these questions uncover. We want you to use these questions to dig deeper into the issues surrounding your data. You may come up with additional questions of your own. We strongly encourage you to share your questions with the rest of your team, so that you are aligned. Data Heads demonstrate their ability to cut through data by setting an example and asking the tough questions on an ongoing basis.

---

[10] Statisticians think a lot about the right sample size to collect, something called power that we'll discuss in Chapter 7.

CHAPTER 5

Explore the Data

If you tell a data scientist to go on a fishing expedition . . . then you deserve what you get, which is a bad analysis.[1]

Thomas C. Redman, "the Data Doc" and Harvard Business Review contributor

Data projects are never as simple as they appear in a boardroom presentation. Stakeholders typically see a polished PowerPoint presentation that follows a rigid script from question to data to answer. What's lost in that story, however, are all the ideas that didn't make the cut: the important decisions and assumptions the data team made along the way to arrive at their answer. A good data team does not follow a linear path but a meandering one, adapting to discoveries in the data. As they get further along in their journey, they circle back to earlier ideas, only to see multiple paths open as a result.

This process of iteration, discovery, and data scrutiny is known as *exploratory data analysis* (EDA). It was formulated by statistician John Tukey in the 1970s as a way to make sense of data with summary statistics and visualizations before applying more complex methods.[2] Tukey saw EDA as detective work. Clues are hidden in data, and the right exploration would reveal next steps to follow. Indeed, EDA is another way to "argue" with your data. It's a

---

[1] Quoted in "Understand Regression Analysis" by Amy Gallo, chapter 10 in HBR Guide to Data Analytics Basics for Managers (HBR Guide Series)
[2] Tukey, J. W. (1977). *Exploratory data analysis* (Vol. 2, pp. 131–160).

fundamental part of all data work that both sets and changes the direction of a project based on what's uncovered.

EXPLORATORY DATA ANALYSIS AND YOU

Exploratory data analysis can be an uncomfortable thought for some: it exposes the subjective nature (the art?) behind all data work. Two teams, given the same problem and data, might take separate analysis paths, possibly landing on the same conclusion. Possibly not. There are just too many decisions to make along the way for any two teams (or individuals) to do everything the same. Each person will bring their own background, ideas, and tools to make recommendations on how best to solve the problem.

Therefore, in this chapter, we present EDA as an ongoing process that is every Data Head's responsibility, whether you're the hands-on data worker or the business leader in the boardroom. You'll learn questions to ask and things to watch out for as you explore data.

Are You a Manager or Leader?

If you are a stakeholder, manager, or subject matter expert, make yourself available to your data team, to the extent possible. Have an open dialogue and expect iteration. Work with them to set correct assumptions. Don't let the data team go on a fishing expedition without the right business context. Otherwise, they may go down paths that make more sense statistically than practically. One wrong assumption could jeopardize everything that follows.

We fully understand that managers can't be as involved in the subtleties of a project as much as data workers. But there's room for a little improvement. You don't need to micromanage. You just can't ignore.[3]

EMBRACING THE EXPLORATORY MINDSET

Dozens of tools and programming languages can help data teams quickly and inexpensively explore their data with summary statistics and visualizations. But EDA should not be thought of as a list of tools or a checklist. It's more of a

---

[3] Stakeholders, to be clear, should not micromanage. There needs to be a level of trust between the business and data teams.

mentality woven into each phase of data work that you can take part of, even without an analytics background.

Questions to Guide You

To help you embrace the exploratory mindset, we'll walk you through a quick scenario against the backdrop of a popular dataset compiled for educational purposes: the Ames Housing Data.[4] This is a glimpse into an EDA process.

Although there's no one right path to follow, there are questions you can ask to help guide the team to a meaningful conclusion:

- Can the data answer the question?
- Did you discover any relationships?
- Did you find new opportunities in the data?

Let's set up the scenario and then go through each of the three questions, discuss why they're worth asking, and share challenges you might face.

The Setup

You work for a real estate start-up company and need to drive traffic to your site. But it's hard to pull visitors from real estate tech giants like U.S.-based Zillow.com. Its famous home-value estimation tool, the Zestimate®, brings people (and profits) to Zillow's brand.[5] To compete, your company needs its own prediction tool. So, you're tasked to build a *model* that takes a home's information as *inputs* and produces an estimated sales price as an *output*.

The boss sends you a dataset to get started. It has 80 columns describing several aspects of hundreds of residential homes that were sold in Ames, Iowa from 2006 to 2011.

Receiving this much data can be overwhelming. However, using the questions outlined previously can help you narrow down how to begin working with the data.

Let's go through them.

---

[4] De Cock, D. (2011). Ames, Iowa: Alternative to the Boston housing data as an end of semester regression project. *Journal of Statistics Education*, *19*(3). You can download the data from www.kaggle.com/c/house-prices-advanced-regression-techniques.

[5] Zillow takes its Zestimate® very seriously. In 2019, it awarded $1 million to a team of data scientists for improving the accuracy of Zestimate® predictions. venturebeat .com/2019/01/30/zillow-awards-1-million-to-team-that-reduced-home-valuation-algorithm-error-to-below-4

CAN THE DATA ANSWER THE QUESTION?

As tempting as it may be to feed your data into the algorithm trend of the moment (e.g., deep learning, covered in Chapter 12), you first need to ask: "Can the data answer the question?" And the answer is often found simply by looking at the data.

Set Expectations and Use Common Sense

You should have a pretty good idea about what information is needed to make an estimate for a home's sales price: size, number of bedrooms, number of bathrooms, the year it was built, etc. These are the popular features home buyers search for on your website. Without them, predicting sales price wouldn't seem reasonable.

You can spot the column names and data types when you open the file. The commonsense features you expect are present, as well as helpful ordinal data (Overall quality of the home, 1–10, 10 being "Very Excellent"), nominal data (Neighborhood), and a host of other features. So far, the data passes an initial sniff test.

Next, you'd likely examine the values the variables take on. Do they cover the scenarios you want to analyze? For example, if you discover the variable "Building Type: Type of Dwelling" only includes single family homes but no apartments, duplexes, or condos, then your model will have limited scope compared to Zillow's. The Zestimate® can predict the sales price of a condo, but if you don't have historical condo data, your company can't reliably predict their sales price.

The lesson: Avoid the fishing expeditions you were warned about in the quote to start this chapter. Make sure the data makes sense for what it's being asked to do.

Do the Values Make Intuitive Sense?

Software will generate a slew of summary statistics for you. Your job is to put the data into context. Check if the summary statistics match your intuitive understanding of the problem. Visualizations are also a key component of EDA—use them to spot anomalies and other weirdness in the data.

Data Visualization Refresher

Let's walk through some quick EDA examples with histograms, box plots, bar charts, and scatter plots. Feel free to skip this if you're comfortable with these figures and the insight they can provide.

You can learn how continuous numeric data is shaped, or distributed, by looking at histograms. Consider the histogram of the sales prices shown in Figure 5.1. There are, for example, about 125 homes in the $200,000 range, and a long tail to the right showing the most expensive homes. That tail pulls the mean sales price ($181,000) past the median price ($163,000). A handful of expensive homes makes the mean larger than the median value.

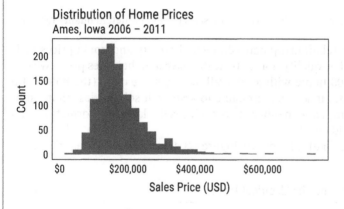

FIGURE 5.1   A histogram showing the shape of sales price

Histograms are helpful to spot anomalies. If you saw negative values (getting paid to buy a house?) or bins with huge counts on the far side of Figure 5.1, which often happens when data is capped (e.g., any value over $500,000 is entered as $500,000), you might want to ask some questions.

Box plots[6] can be used to compare data across several groups. Figure 5.2 shows a box plot for each quality rating of a home: 1 being poor and 10 very excellent.

---

[6] Box plots are also called box-and-whisker plots. The "box" contains the middle half of the data (values between the 25th and 75th percentiles), the line in the box is the median, and the "whiskers" show the range of the remaining data points. The dots beyond the whiskers are potential outliers.

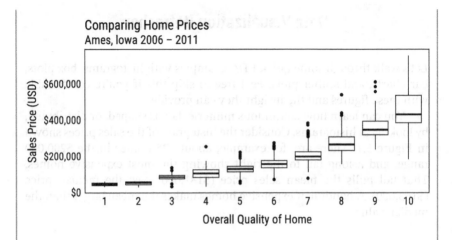

FIGURE 5.2   Using box plots to compare sales prices at different quality rankings

Here, the relationship between overall quality and home prices feels intuitive. Higher-quality homes typically have a higher sales price. We can spot a $200,000 home with an overall quality score of 10 (the bottom tip of the line), but it seems reasonable to assume it sold for less than other perfect-10 homes due to other factors. This is the kind of information data workers should check.

Bar charts, like that shown in Figure 5.3, show counts of categorical data.

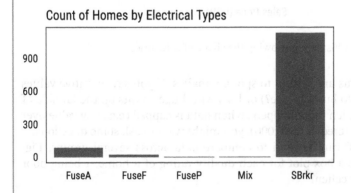

FIGURE 5.3   A bar chart showing the counts by types of electrical installation

Not all visuals will be interesting at first glance. However, it's still a good idea to view such visualizations if only to reinforce (or perhaps challenge) the previous question—does the data make intuitive sense? Figure 5.3, after all, shows almost all homes have the same value for this feature. For your task, however, this information is helpful. Since most homes have the same value for this variable, it likely won't contribute to any meaningful differences in the sales price of homes.

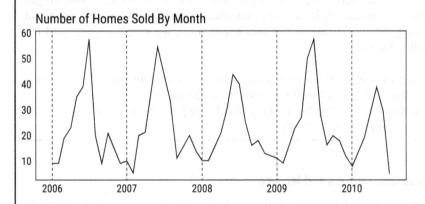

Number of Homes Sold By Month

FIGURE 5.4 A line chart showing the number of houses sold in different months

Figure 5.4 shows a line chart with the number of houses sold per month in the data. You can easily visualize a phenomenon where home sales spike in the summer and retract in the winter—an example of *seasonality*. Line charts are helpful to spot such trends.

Home Size and Sales Price

FIGURE 5.5 A scatter plot showing square footage and sales price

Next, we can examine a scatter plot showing houses plotted based on their size (square footage on the first floor) and sales price (see Figure 5.5).

Figure 5.5 shows an intuitive pattern. Larger homes generally sell for more money. Of course, the rule isn't always true. Sometimes small homes cost more than large homes. There's always variation, but the overall trend is there. And since we are trying to predict sales price as an output, square footage seems like a great piece of information to have.

This section was just a taste of what kind of information and insight you can get quickly by plotting your data. If you'd like to learn more about using data visualization for data exploration, we would recommend the following titles:

- *Now You See it: Simple Visualization Techniques for Quantitative Analysis,* by Stephen Few (Analytics Press, 2009)
- *The Visual Display of Quantitative Information,* by Edward Tufte (Graphics Press, 2011)

Watch Out: Outliers and Missing Values

Every dataset will have anomalies, outliers, and missing values. How you deal with these matters.

For instance, the box plots in Figure 5.2 used a rule-of-thumb to flag several data points as potential outliers. But just because a data graphic classifies certain points as "outliers," don't turn off critical thinking and automatically delete these points assuming they can't be useful. You'll never catch Zillow removing useful information from its datasets simply because a visualization described these as outliers. Use the context of the data—homes that cost much more than the bulk of other homes is both a known and a common feature of real estate data. Recall the lessons from the previous chapter. You should at least have good business justification to remove outliers. Do you have one here?

And what about missing values? Does a missing value in "Basement Size" mean the house has a basement and the area is unknown? Or does it mean there is no basement and the value should be 0?

If we are diving into the weeds a little, that's our intent. Data workers make hundreds of these tiny decisions during projects. The cumulative effect can be substantial. Left to their own devices—and without the guidance of domain expertise—data workers may continue chipping away at the data, removing difficult and nuanced cases, until the data is too detached from the reality it's trying to capture to be useful. This is why it's important for everyone, including managers, to really understand what their data teams are doing.

DID YOU DISCOVER ANY RELATIONSHIPS?

Fortunately for us, a first pass of the housing data with summary statistics and visualizations seems encouraging and you think the data can indeed be used to build a predictive model for sales price, so you press on to the next question: "Did you discover any relationships?"

Visualizing the data has given you a head start: higher overall quality and larger square footage are unsurprisingly related to higher sales prices. This is the feedback you want from data. The relationships make sense and the variables you've plotted will help you build a model to predict sales price. What other variables share a relationship with sales price?

At this point, summary statistics can help steer you toward interesting patterns and relationships in the data because generating every possible scatter plot may not be practical. Instead, the relationship found in scatter plots can reduce down to the summary statistic *correlation*, which is suggestive (but not proof) of a relationship between two numeric variables.

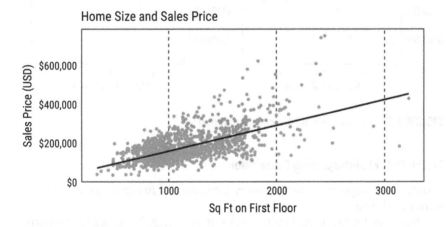

FIGURE 5.6 Square footage and sales price have a correlation of 0.62, which measures the tightness of the data points around the solid linear trend line.

Understanding Correlation

Correlation is a measure of how two variables are related. The most common type of correlation used in business is the *Pearson correlation coefficient*, a statistic between −1 and 1 that measures the linear relationship (think simple straight lines) between pairs of numbers shown on a scatter plot. Correlation can be positive, meaning an increase in one variable is associated with an increase in the other: larger homes sell for more money. Or, correlation can be negative: heavier cars get worse gas mileage. For a visual reference, the correlation of home size and sales price, shown in Figure 5.6, is 0.62. The "tighter" the points around a linear trend, the higher the correlation.[7]

Correlation can help here in two ways. First, finding variables correlated with sales price would help predict it. Second, correlation can help reduce redundancies in your data because two highly correlated variables contain roughly the same information. Imagine two columns in your data: the area of the home in square feet and the area in square meters. These two are perfectly correlated; only one is needed in an analysis.

While most of us have a basic grasp of correlation and report the metric often, it can deceive. Let's review how.

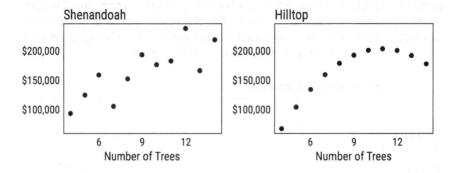

FIGURE 5.7   Two datasets with a correlation of 0.8

Watch Out: Misinterpreting Correlation

People often forget that correlation is a measure of *linear* trend, and not all trends are linear.

Suppose, for instance, you're analyzing two neighborhoods in the housing dataset, each with 11 homes. Crunching some statistics reveals that the number of trees on a property is highly correlated with sales price within these neighborhoods. The correlation is a strong 0.8: properties with more trees tend to sell for more money.

But a visual check of the data exposes something unexpected. In Figure 5.7, the data for the neighborhood on the left shows what we'd typically expect to see with a high correlation: a linear trend with data points scattered about. But the plot on the right shows the number of trees is associated with an increase in sales prices *only up to a point* (11 trees). After that, it trends

---

[7] Correlation does not mean "steepness." Two perfectly correlated variables could appear almost flat (though not exactly horizontal).

downward. In the Hilltop community, some properties might have too many trees crowding their lawn.

Full disclosure: the data shown in Figure 5.7 did not come from the Ames dataset we've been exploring but from a popular dataset called Anscombe's Quartet,[8] four datasets with identical summary statistics but clearly different visualizations. (We are showing just two and adjusted the data to reflect the real estate theme.)

The lesson: Use visualizations to verify noteworthy correlations in your data because the linear trend that correlation can identify may not tell the full story.

Not Correlated but Still Interesting

Figure 5.8 shows two plots, each with identical, near zero correlation coefficients. Don't let that trick you into thinking there's nothing funny going on. And while you won't run into many *data*sauruses like in the left plot, you might come across the scenario in the right plot: five groups of linearly correlated data in fact, but when viewed as a single group, it's not linearly correlated at all. This is known as *Simpson's Paradox*, a topic you'll learn more about in Chapter 13.

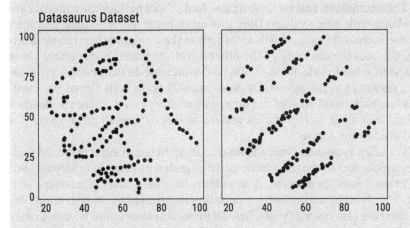

FIGURE 5.8 Datasaurus: Data is free to download and explore.[9] Like Anscombe's Quartet, both datasets shown here have identical summary statistics.

---

[8] Anscombe, F. J. (1973). Graphs in statistical analysis. *The American Statistician, 27*(1), 17–21. We multiplied the dependent variable by 22,000 to create reasonable sales price examples.

[9] The datasaurus was created by Alberto Cairo and the data is available on GitHub: github.com/lockedata/datasauRus

Watch Out: Correlation Does Not Imply Causation

Chances are, you've heard the phrase "Correlation does not imply causation" before.[10] But it's worth reiterating because of how much it's ignored and even misunderstood.

When two variables are correlated, even strongly correlated, it does not mean that one is causing the other. Yet people fall into this trap far too often, seeking to build a narrative whenever two variables move together. There are typical silly examples statisticians use to show correlation does not prove causation: Ice cream sales are correlated with shark attacks (both spike in the summer months). Shoe size is correlated with reading ability (both increase over time). But to suggest that reducing ice cream sales will mitigate shark attacks, or that buying bigger shoes helps you read, is clearly a joke. There are other factors at play—outside temperature in the ice cream example, age in the shoe size example—that obviously play a part in the spurious relationships.

But when the correlations aren't built around jokes and the true causal factor isn't clear, the mantra "correlation does not imply causation" is often forgotten.

For example, in real estate data, you find school performance metrics are correlated with home values. Does this mean better schools cause a home's value to increase? Good schools seemingly make a neighborhood more desirable. Or, does the causality go the other direction: higher home prices cause a boost in school performance? Maybe the increase in tax revenue provides more resources to the school. Or does causality go in both directions, creating a feedback loop? Most of the time, we just don't know. There are clearly a multitude of other factors at play, and it'd be rare to have all the answers you need inside your dataset.

It's safer to assume "there is no causality" between two correlated variables unless someone has conducted an experiment proving otherwise. But don't take this to the extreme. Both authors have seen cases in business, university, and media settings where causation is assumed when it shouldn't be. But there are also cases as well where an important association is immediately dismissed as being an assumed causation fallacy. (See the sidebar for an example of causality being dismissed when it should not have been.)

---

[10] Your authors debated if it's even possible to not mention "Correlation Does Not Imply Causation" in a data book. You can see the outcome.

Smoking and Lung Cancer

Ronald A. Fisher, who is considered one of the foremost statisticians of the 20th century, who even developed and contributed to the techniques described in this book, was largely skeptical of studies linking tobacco use to cancer.

Fisher was most concerned with confounding variables. What if, for example, some people were genetically predisposed to get lung cancer and wanted to smoke to alleviate their symptoms? According to Fisher, early studies on the risks of tobacco use had committed "an error. . .of an old kind, in arguing from correlation to causation."[11]

And yet, we now know the link between the two is indisputable. Inasmuch as we should be wary against seeing causality where there is none, we should also be careful not to dismiss relationships that haven't proven to be causal *yet*.

DID YOU FIND NEW OPPORTUNITIES IN THE DATA?

EDA is not just a process to better understand data and set a path forward to solve problems. It's also a chance to find other opportunities in the data; problems that might be valuable to your organization. A data scientist may spot something interesting or weird in a dataset and then formulate a problem.

However, you don't know if anyone needs the solution you've found until you follow the steps in Chapter 1, "What Is the Problem?"

CHAPTER SUMMARY

To become a Data Head, you need to embrace an ongoing process of exploratory data analysis. This will allow for:

- A clearer path forward to solve the problem.
- Refining the original business problem, given the constraints identified in the data.

---

[11] Fisher, R. A. (1958). Cancer and smoking. *Nature, 182* (4635), 596.

- Identifying new problems to solve with the data.
- Cancellation of the project. While unsatisfying, EDA is successful if it stops you from wasting time and money on a dead-end problem.

We guided you through the process using a real estate dataset (one we'll return to in Chapter 9 to finally build that model we've been talking about) and talked about common hurdles you may encounter.

The flow of this chapter assumes you can be a part of the EDA process from beginning to end. There will be times when this isn't possible, particularly for senior leaders overseeing many projects. But missing the early stages does not absolve Data Heads of their responsibility to have an exploratory mindset. If you come into a project near the end, ask why the data team chose the analysis they did and what challenges they faced. You may uncover assumptions you would not have made yourself.

CHAPTER 6

Examine the Probabilities

Many people's notion of probability is so impoverished that it admits [one] of only two values: 50-50 and 99%, tossup or essentially certain."

—John Allen Paulos, mathematician and author of "Innumeracy: Mathematical Illiteracy and Its Consequences"[1]

Let's talk about probability—the language of uncertainty—and rekindle the conversation we started in Chapter 3, "Prepare to Think Statistically." To recap, there's variation in all things; variation creates uncertainty; and probability and statistics are tools to help us manage uncertainty.

Our all-too-brief section on probability ended with a message: *be mindful and recognize that your intuition can play tricks on you.*

A fair statement, but such topics like probability deserve more than a bumper sticker warning. Complete understanding, if there is such a thing, requires reading massive textbooks, listening to long lectures, and a lifetime of study and debate. Even then, experts disagree on the interpretation and philosophy of probability.[2] You might not have the time or desire for that level of discussion (neither do we), so we'll spare you those debates and keep the focus on what you need to know to keep your intuition sharp and be successful in your work.

Our goal in this chapter, then, is to move you beyond a bumper sticker grasp of probability and to teach you some of its language, notation, tools,

---

[1] Paulos, J. A. (1989). *Innumeracy: Mathematical Illiteracy and Its Consequences* (1st ed.). Hill and Wang.
[2] Search online for "Probability Interpretations" to see what we're talking about.

and traps. By the end of this chapter, you will be able to think and speak about probabilities in the workplace, even if you are not the one calculating the numbers—and you will be able to ask hard questions of probabilities presented to you. Indeed, the willingness to wade into discussions about probability and uncertainty is an important step in your development as a Data Head.

TAKE A GUESS

Here's a thought exercise to get you started.

Your Fortune 500 company was the victim of a cyber-attack, and hackers infected 1% of all laptop computers with a virus. The valiant IT team quickly creates a way to test if a laptop is infected. It's a good, almost perfect test. In fact, the IT team's research shows that if the laptop has the virus, the test will be positive 99% of the time. And if a laptop does not have the virus, the test will come back negative 99% of the time.

The team finally tests your laptop for the virus—the result, positive. The question is, what is the probability your laptop actually has the virus?

Take a moment to think of an answer before moving on.

The correct answer is 50%. (We'll prove it later in the chapter.)

Surprised? Most people are.

The answer isn't intuitive. In fact, you knew probability could play tricks on your mind, and it *still* did. That's the frustrating thing about probability—every problem is a brainteaser. Don't feel discouraged, however, if you got the wrong answer. The real test is whether you paused to think about how uncertain you were before peeking at the answer. Did you?

Not everyone does. The truth is that most people don't have a grasp of, or respect for, probability. Want proof? People still buy scratch-off lottery tickets, flock to Las Vegas, and buy the extended warranty on their television. They are content being woefully ignorant about probability—especially when decisions are tied to a potential payoff (slot machines) or avoiding perceived future frustration (television warranties). Luckily, this chapter will give you a firm grasp of probability, its rules, and its misconceptions.

Let's get started.

THE RULES OF THE GAME

Probability lets us quantify the likelihood that an event will occur.

Before we introduce the math, it's worth considering that our brains are wired for probabilities, in that we use probabilistic statements all the time in daily life. For any event in your life, you don't know for certain exactly what will happen, but you do know that some outcomes are more likely than others. For example, you might hear around the office:

- "It's *highly probable* they'll sign the contract!"
- "There's a *small chance* we'll miss next Monday's deadline."
- "It's *doubtful* we'll hit our quarterly goals."
- "Trevor *usually* shows up late to meetings."
- "The weather channel says it's *likely* to rain today—let's move the company retreat."

But these everyday probability terms are mushy enough to seem correct, even when they're not. Indeed, two people might have different perceptions of how often a "highly probable" or "likely" event occurs—which means everyday language isn't going to cut it. We need to use numbers, data, and notation to quantify probability statements so that what we say is more than just a gut feeling (even if our gut feelings have a high degree of reliability). Not only that, we need to abide by certain rules and logic in probability.

Notation

Probability, as we said earlier, lets us quantify the likelihood that an event will occur. An event can be any outcome, from basic (flipping heads on a coin) to the complex ("Donald Trump will win the 2016 election"). A child can understand the 50-50 chance of flipping a coin, but the entirety of the polling and forecasting industry struggled to predict the 2016 election, even after analyzing terabytes of data.

We'll focus on the simple cases for this quick lesson.

Probability is measured by a number between 0 and 1, inclusive, with 0 being impossible (rolling a 7 on a six-sided die numbered 1–6) and 1 being certain (rolling a number less than 7 on a six-sided die). It's often expressed as a simple fraction (flipping heads on a coin has probability 1/2), or as a percentage (you have a 25% chance of picking a "spade" from a standard deck of playing cards). Many people use all of these—numbers, fractions, and percentages—interchangeably.

To save space, we use shorthand and call probability P. Descriptions of events are also condensed. For example, "The probability of flipping a fair Coin on Heads equals 1/2" can be shorted to $P(C == H) = 1/2$. Or, even shorter, $P(H) = 1/2$. In fact, the entire previous paragraph can be written as described in Table 6.1.

TABLE 6.1 Probabilities Scenarios with Associated Notation

Scenario	Notation
Probability of rolling a 7 on a six-sided die	$P(D == 7) = 0$
Probability of rolling less than a 7 on a six-sided die	$P(D < 7) = 1$
Probability of picking a spade from a card deck	$P(S) = 0.25$

Using "==" Instead of "="

If you've taken a probability or stats class, the probability notation likely isn't new; however, we're adding a twist here that we hope improves clarity.

Notice that we are testing the probability of a coin flip being heads, so we say $P(C == H)$ instead of $P(C = H)$. The reason we do this is to differentiate between the two sets of equals in our equation. When we write ==, we're effectively testing what the coin flip result, C, equals.

On the other hand, when we write $P(C == H) = 1/2$, the single equals at the end of the notation indicates that the result of $P(C == H)$ is equal to 1/2.

This notation follows the Boolean logic syntax in many programming languages.

The notation $P(D < 7) = 1$ is expressing a *cumulative probability*—a range of outcomes. It's saying, "The probability you roll a number less than 7 on a Die is 1." It's adding $P(D == 1) + P(D == 2) + P(D == 3) + P(D == 4) + P(D == 5) + P(D == 6) = 6 \times 1/6 = 1$ (see Table 6.2). The sum of all possible outcomes must equal one.

TABLE 6.2 Cumulative Probability of a Die Roll Less than 7

Scenario	Notation	Probability
You roll a 1	$P(D == 1)$	1 / 6
You roll a 2	$P(D == 2)$	1 / 6
You roll a 3	$P(D == 3)$	1 / 6
You roll a 4	$P(D == 4)$	1 / 6
You roll a 5	$P(D == 5)$	1 / 6
You roll a 6	$P(D == 6)$	1 / 6
You roll a number less than 7	$P(D < 7)$	$6 / 6 = 1 = 100\%$

Conditional Probability and Independent Events

When the probability of an event depends on some other event, it's called a *conditional probability* and uses notation called a *vertical bar*, |, read as "given." A few examples will make this clearer:

- The probability Alex is late to work is 5%. $P(A) = 5\%$.
- The probability Alex is late to work *given* he has a flat tire is 100%. $P(A \mid F) = 100\%$.
- The probability Alex is late to work *given* there's a traffic jam on Interstate 75 is 50%. $P(A \mid T) = 50\%$.

As you can see, the probability of an event occurring depends heavily on the event, or events, preceding it.

When the probability of an event does not depend on some other event, the two events are *independent*. For example, the conditional probability of drawing a spade given a coin flipped heads, $P(S \mid H)$ is the same as the probability of drawing a spade by itself, $P(S)$. In short, $P(S \mid H) = P(S)$ and, for good measure, $P(H \mid S) = P(H)$, because there is no dependency between the two events. The deck of cards doesn't care what happened with the coin and vice versa.

The Probability of Multiple Events

When modeling the probability of multiple events occurring, the notation and rules depend on how the multiple events occur—whether two events occur together (it floods and the power goes out), or whether one event or another occurs (either it floods or the power goes out).

Two Things That Happen Together

To start, we'll talk about two events happening at the same time.

P(flipping a heads on a coin) $= P(H) = 1/2$.

P(drawing a spade from a deck of cards) $= P(S) = 13/52 = 1/4$.

The probability of both happening, flipping a heads *and* drawing a spade can be denoted $P(H, S)$, with the comma representing "and."

In this case, the events are *independent*. One event has no impact on the other. When events are independent, you can multiply the probabilities: $P(H, S) = P(H) \times P(S) = 1/2 \times 1/4 = 1/8 = 12.5\%$. Seems simple enough.

Now consider a slightly harder example. Recall, the probability Alex is late to work is 5%, $P(A) = 5\%$. The probability Jordan being late to work, however, is 10%. $P(J) = 10\%$. What can you say about the probability of both of us being late to work, $P(A, J)$? For context, we live in different states, Alex has a 9–5 job, and Jordan works for himself.[3]

The first guess is $P(A, J) = P(A) \times P(J) = 5\% \times 10\% = 0.5\%$. A seemingly rare event, but are these two events truly independent? It might seem that way at first because we live and work in separate places. But no, the events are not independent. We are, after all, writing a book together. Both of us might be running late because we were up late arguing how best to explain probability. So, the probability Alex is late depends on whether Jordan is running late too. A conditional probability therefore is needed. Let's assume the probability Alex is late given Jordan is late is 20%, which we write as $P(A \mid J) = 20\%$.

This brings us to the true formula for two events happening at the same time, called the *multiplicative rule*. It can be written as $P(A, J) = P(J) \times P(A \mid J) = 10\% \times 20\% = 2\%$. Said in words, the probability that Alex and Jordan are both late equals the probability of Jordan being late multiplied by the probability of Alex being late, *given* Jordan is late.

The final probability, 2%, can never be more likely than the smallest of the individual probabilities of either one of us being late, $P(A)$ and $P(J)$, which is 5% for Alex. That's because Alex has a 5% chance over all possible scenarios, including the scenarios when Jordan is late.

That brings us to an important rule in probability: *The chance of any two events happening together cannot be greater than either event happening by itself.*

Figure 6.1 shows this rule using Venn diagrams. If you think of probability as area, you'll see how the intersection, or overlap, of the circles (events) A and J can never be larger than the smallest circle.

One Thing or the Other

What about one event *or* some other event occurring at the same time? As with all lessons in statistics and probability, it depends. Start with a guess and adjust with the evidence provided.

When the events cannot occur at the same time, it's an easy addition problem. You can't roll both a 1 and a 2 on a die at the same time, so the probability of rolling a 1 *or* a 2 is $P(D == 1 \text{ or } D == 2) = P(D == 1) + P(D == 2) = 1/6 + 1/6 = 2/6 = 1/3$.

---

[3] Can you really be late if you work for yourself? For this example, yes.

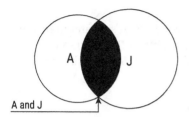

FIGURE 6.1 Venn diagram showing the probability of two events happening together cannot be greater than either event by itself.

Let's return to your truant authors for a slightly harder example. Instead of asking when Alex *and* Jordan will both be late to work, what about the probability that Alex *or* Jordan will be late? This is denoted $P(A$ or $J)$.

Here's what you know. $P(A) = 5\%$ and $P(J) = 10\%$. A first guess might be $P(A) + P(J) = 15\%$, a reasonable start. Out of 100 days, Alex will be late 5 days and Jordan will be late 10. Add them up to get 15 days, or 15% of 100. If the events were *mutually exclusive* and *never occurred together*, this would be correct.

Remember, however, that there's overlap when we both could be late. Figure 6.1 shows this. We are sometimes late to work because of each other, which is to say the probability that Alex is late and Jordan is late, $P(A, J)$, is more than 0. We can't just add both probabilities, because that would double count the days when we were both late. To compensate, we have to subtract the probability when we're both stumbling into work after a late-night writing session, which was $P(A, J) = 2\%$. That's 2 days out of 100 where we overlapped late days, giving the final count as 5 for Alex, 10 for Jordan, minus the 2 days when they overlapped: $5 + 10 - 2 = 13$ and $13/100 = 13\%$.

With that information, we can formally give the additive rule for when one event or another happens at the same time: $P(A$ or $J) = P(A) + P(J) - P(A, J) = 5\% + 10\% - 2\% = 13\%$.

Remember the Overlap

The challenging part for some when dealing with the probability of multiple events is subtracting the overlap. Realize, however, that this must take place because probability cannot ever be more than 1. Let's use dice because it's easier to comprehend. The probability of rolling a number greater than 2 is 4/6. The probability of rolling an odd number is 3/6. If you want to know the probability of one event or the other happening, you cannot add because you'd have $4/6 + 3/6 = 7/6 = 1\,1/6$. It's greater than one and violates the rules of probability. We must subtract the overlap when a die is greater than 2 *and* odd—i.e., the numbers 3 and 5, which have probability 2/6.

> **Problem statement:** $P\big(D > 2 \text{ or } D \text{ is odd}\big) =$
>
> **Additive rule:** $P\big(D > 2\big) + P\big(D \text{ is odd}\big) - P\big(D > 2, D \text{ is odd}\big) =$
>
> **Plug-in probabilities:** $4/6 + 3/6 - 2/6 =$
>
> **Answer:** 5/6
>
> Rolling a 2 is the only number that doesn't satisfy either condition.

That was a lot of notation, dice, coin flips, and your authors running late to work. So, to take a step back from notation and numbers, let's do a thought exercise without either.

PROBABILITY THOUGHT EXERCISE

Sam is 29 years old, reserved, and very bright. He majored in economics in his home state of California. As a student, he was obsessed with data, volunteered at the university's statistical consulting center, and taught himself how to program in Python (a computer programming language).
Which of these is more probable?

1. Sam lives in Ohio.
2. Sam lives in Ohio *and* works as a data scientist.

The correct answer is #1, even though nothing in the description suggests Sam would live in Ohio *without* also being a data scientist. This is a rebranding of the popular Linda problem in the book, *Thinking, Fast and Slow*[4], and it's a problem most people get wrong. How did you do?
Did you pick #2? Perhaps it was because we gave you background into the fact that Sam knew programing and *could be* a data scientist. #2 seems more probable precisely because it mentions an event aligned with Sam's background, but it's still less probable than number #1.
Here's why. This example is stripped of notation and numbers, but it still reflects an important lesson in the previous section. *The chance of any two*

---

[4] Kahneman, D. (2013). *Thinking, Fast and Slow* (1st ed.). Farrar, Straus and Giroux.

events happening together can't be greater than either event happening by itself. The more "and" qualifications you add to any statement, the more you narrow the possibilities. For Sam to be a data scientist *and* live in Ohio, he must first live in Ohio. He might live in Ohio and work as an actuary.

Remember, the probability of two events happening together uses the multiplicative rule. The probability of Sam living in Ohio *and* working as a Data scientists can be represented as $P(O, D) = P(O) \times P(D \mid O)$. And because probabilities are no more than one, multiplying $P(O)$, the probability Sam lives in Ohio, by *any other probability* can never make the resulting number, $P(O) \times P(D \mid O)$, increase. Thus, it's impossible for $P(O, D)$ to ever be greater than $P(O)$, no matter how right it felt in the moment to guess #2.

Still wrestling with it? You may have read #2 as a conditional probability: what is the probability Sam lives in Ohio *given* he works as a data scientist, $P(O \mid D)$. This *can* be more probable than Sam living in Ohio, $P(O)$; however, the language difference between "*and*" and "*given*" matters.

Take an easier example: The New York Yankees baseball team has loyal fans across the world. Suppose there's a game right now with millions watching, both at the stadium and on television. Now randomly select a person in the world. With billions of people in the world, it's very unlikely you'd pick a Yankees fan. It's even more unlikely you'd select a Yankees fan sitting at the game because not all fans can attend the game. But if you were *given* the ability to randomly select a person at the game, everything changes. It's highly probable they are a Yankees fan.[5]

Thus, the probability of a person being a Yankees fan *and* attending the game is vastly different from the probability of a person being a Yankees fan *given* they are at the game.

Next Steps

After that thought exercise, it's worth stating the warning we shared at the beginning of the chapter: *be mindful and recognize that your intuition can play tricks on you.* Probabilities will continue to confound and confuse. Perhaps the best we can do to combat this is point out some common traps.

In that spirit, and now that you're armed with the rules and notation of probability, we'll dedicate the rest of this chapter with sections to help you be aware and think critically about the probabilities you examine in your work. Here are some guidelines to keep you on the right track:

- Be careful assuming independence
- Know that all probabilities are conditional
- Ensure the probabilities have meaning

---

[5] Not 100% because the opposing team will also have fans at the game.

BE CAREFUL ASSUMING INDEPENDENCE

If events are independent, you can multiply their probabilities together: the probability of flipping two heads in a row on a fair coin is $P(H) \times P(H) = 1/2 \times 1/2 = 1/4$. But not all events are independent, and you need to be careful with this assumption when you're calculating or reviewing probabilities.

We mentioned this early in the book with the mortgage crisis of 2008. The probability of someone defaulting on their mortgage is not independent from the probability their neighbor defaults, though for many years Wall Street didn't think so. Both are intrinsically tied to the overall economy and state of the world.

Yet, assuming independence when events are not is a mistake made again and again. Your company might be making this mistake in strategy sessions. And this risk is underestimating, perhaps grossly underestimating, the probability multiple events can occur simultaneously.

Consider, for example, a C-level strategy session in a boardroom where they're placing three big bets for the company in the coming year—high-profile, exciting, but risky projects. Call the projects A, B, C. The execs know it's possible each project could fail, and estimates the probability of failure for each project as $P(A \text{ fails}) = 50\%$, $P(B \text{ fails}) = 25\%$, and $P(C \text{ fails}) = 10\%$.

Someone grabs a calculator and multiplies the probabilities: $50\% \times 25\% \times 10\% = 1.25\%$. The execs are ecstatic—there is only a 1.25% chance all three fail. These are big bets, after all. Just one successful project will justify their investment in the three. And, because all outcomes must sum to 1, the probability *at least one* project will succeed would be 1 minus the probability *all* fail, or $1 - 0.0125 = .9875 = 98.75\%$. Wow, they think, almost a 99% probability of overall success!

Alas, their math is wrong. The events are dependent on the overall success of the company, which could be brought down by a host of examples: corporate scandal, poor quarterly results, or some larger event tied to the health of the global economy, like the COVID-19 pandemic. The events A, B, and C are dependent on several factors. Therefore, by wrongfully assuming independence, they are underestimating the probability all projects will fail next year, and thus overestimating the chances at least one will be successful.

Lest you think that doesn't seem important, remember the 2008 financial collapse and the ensuing recession.

Don't Fall for the Gambler's Fallacy

On the flip side, some events are independent but not viewed that way. This creates a different kind of risk that casinos thrive on, where people overestimate how likely something can occur based on recent events.

A fair coin, even if it lands on 10 heads in 10 consecutive flips, still has $P(H) = 50\%$ on the next flip. With independent events, your chances for an event *do not increase or decrease* based on past performance. But gamblers falsely believe the probabilities change—hence, the name Gambler's fallacy.[6]

Every roll of the dice is independent of the previous roll, each pull on a slot machine has no memory of the pull before, and each spin on the roulette wheel does not depend on the last. Yet gamblers fall victim to finding patterns in these events. They either think a slot machine is "due" to hit because it hasn't shot out coins in a while, or they believe the dice are "hot"—winning dice keep winning.

But each instance has the same probability of winning as the last. And since it's in a casino, the probabilities are not in your favor. But amateur gamblers bet big when a cluster of rare events occurs, thinking it's their lucky day to get paid. Oh, how wrong they are. But, hey, maybe the casino will give them a "free" breakfast buffet.[7]

ALL PROBABILITIES ARE CONDITIONAL

All probabilities are conditional in some way. A coin flip with $P(H) = 50\%$ is conditional on the coin being fair. Rolling a die with $P(D == 1) = 1/6$ is conditional on using an unloaded die. The probability of success on a data project is conditional on the collective wisdom of the data team, the correctness of data, the difficulty of the problem, whether your computer gets a virus, whether a pandemic shuts down the company, and on and on.

Think, also, about how businesses and people judge success and competence. It's usually based on past successes. Companies hire a consultant with a winning track record or the attorney who wins the most cases, or a person demands the heart surgeon with the smallest patient death rate. The consultant may make his clients' money 90% of the time, the attorney wins 80% of her cases that go to trial, and the heart surgeon's patients have a low 2% mortality rate.

Here's how they could game the probabilities. The consultant, lawyer, and surgeon can decide whether or not to participate. They have a good idea of when they'll be successful, and if the chances don't look favorable, they can say "no." The probability of success for each is dependent on each selecting projects most likely to succeed and avoiding those that might hurt their numbers.[8]

You must think of all the factors behind the probability numbers you see.

---

[6] The belief that past independent events might occur given enough time is also sometimes called the "law of averages," a scientific sounding name for wishful thinking.

[7] The authors have nothing against breakfast buffets.

[8] To be clear, we are not saying consultants or surgeons do this. Just attorneys.

Don't Swap Dependencies

Another trap to sidestep is assuming $P(A \mid B) = P(B \mid A)$ for two events A and B. Notice how the dependencies are swapped: In one case, A depends on B. In the other, B depends on A.

Here's an example showing the two are not equal. Let event A be "Living in New York state" and event B be "Living in New York City." $P(A \mid B)$, the probability of living in New York state *given* that you live in New York City is quite different than $P(B \mid A)$, the probability of living in New York City given you live in New York state. The former is a guarantee, $P(A \mid B)$ = 1; the latter is not, as about 60% of New York state residents live outside New York City.

It's clearly a mistake in an easy example like this, but swapping the dependencies and assuming $P(A \mid B) = P(B \mid A)$ is an error so common it was given a name and a Wikipedia article—Confusion of the Inverse.[9] In fact, it's the very error you might have made in the thought exercise at the beginning of this chapter.

Let's return to that problem now.

Your company was hacked and 1% of laptops have a virus. Event + is a positive test, event − is negative, and event V is infected with the virus. You were given the following information: $P(+ \mid V)$ = 99%, $P(- \mid not\ V)$ = 99%, and $P(V)$ = 1%. In other words, the probability of a positive test given the laptop has the virus is 99%, the probability of a negative test given the laptop does not have the virus is 99%, and the probability a random laptop has the virus is 1%.

We wanted to know the probability a computer had a virus given a positive test result, $P(V \mid +)$, but this is where the "confusion of the inverse" emerged. We asked for $P(V \mid +)$, not $P(+ \mid V)$, yet many people, when presented with the thought exercise, guess a number close to $P(+ \mid V)$ = 99%.

The probabilities $P(V \mid +)$ and $P(+ \mid V)$ are not the same, but they are linked through one of the most famous theorems in all of probability and statistics—Bayes' theorem.

Bayes' Theorem

Bayes' theorem, which dates back to the 1700s, is a clever way to work with conditional probabilities that's been applied everywhere from battles

---

[9] Confusion of the Inverse: en.wikipedia.org/wiki/Confusion_of_the_inverse. Accessed on July 4, 2020.

to finance to DNA decoding.[10] Bayes' theorem states the following for two events, *A* and *B*:

$$P\left(A|B\right)\times P\left(B\right)=P\left(B|A\right)\times P\left(A\right)$$

Let's unpack this a little because the notation can be intimidating. More important than memorizing the formula (or any formula for that matter) is understanding what it's doing and why it's worth knowing.

Bayes' theorem enables us to relate the conditional probability of two events. The probability of event A *given* event B is related to the probability of event B *given* event A. They *are not equal* (that's the "confusion of the inverse") but *related* by the preceding equation.

Why do people care? In practice, one of the conditional probabilities is known and people want to find the other. For example:

- Medical researchers want to know the probability a person has a positive cancer screening test *given* the person has cancer, $P(+ \mid C)$, so they can create more accurate tests to provide treatment right away. Policy makers want to know the inverse—the probability a person has cancer given they test positive on the screening test, $P(C \mid +)$, because they don't want to subject people to unnecessary cancer treatment based on a false positive (when test says "cancer" and no cancer is present).
- Prosecutors want to know the probability a defendant is guilty *given* the evidence, $P(G \mid E)$. This depends on the probability of finding evidence given the person is guilty, $P(E \mid G)$.
- Your email provider wants to know the probability an email is spam given it contains the phrase "Free Money!", $P(Spam \mid Money)$. Using historical data, it can calculate the probability an email contains the phrase "Free Money!" given that it's spam, $P(Money \mid Spam)$. (You'll learn more about this example in Chapter 11.)
- Returning to the thought exercise, you want to know the probability of your computer having a virus given a positive test, $P(V \mid +)$. You know the inverse, the probability of a positive test given a computer has the virus, $P(+ \mid V)$.

All these examples are linked through Bayes' theorem—a slick way to flip the conditional probabilities. That's the good news. The bad news, or rather

---

[10] For a complete and excellent history about Bayes' theorem, check out the book McGrayne, S. B. (2011). *The Theory That Would Not Die: How Bayes' Rule Cracked the Enigma Code, Hunted Down Russian Submarines, and Emerged Triumphant from Two Centuries of Controversy* (American First ed.). Yale University Press.

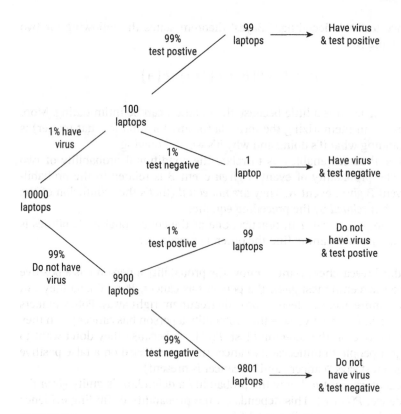

FIGURE 6.2   Tree diagram for scanning computers for a virus at a large company

challenging news, is that the actual calculation of some of the pieces in Bayes' theorem can be a pain. Not all probabilities are easy to come by. For instance, the rate at which a person might have cancer given they've tested positive in a cancer screening test might be easier to find out than the incidence a person has cancer given that they failed the first preliminary test.

One way to assess whether you have enough information to use Bayes[11] is by building a tree diagram (see Figure 6.2). We'll use the thought exercise as an example and finally reveal why the correct answer is 50%. Suppose there are 10,000 laptops in the company. We will use the implied understanding that if 99% of computers with a virus test positive, 1% will test negative, $P(- \mid V) = 1\%$. Likewise, since 99% of laptops without the virus will test negative, 1% of laptops without the virus will test negative, $P(+ \mid not\ V) = 1\%$.

---

[11] Here, we say "using Bayes" as a shorthand for "applying Bayes' Theorem."

Starting with 10,000 laptops, as shown in Figure 6.2, and the information we provided, you can see how they split into the four final groups: laptops with the virus that tested negative or positive and laptops without the virus that test negative or positive. Let's consider what this means. If you look at the tree diagram, you'll notice there are only two possible branches we're interested in. The first case is having the virus and testing positive—that's 99 laptops. The second case is not having the virus but still testing positive—also 99 laptops. These are called *false positives*.

Here's the deal: we already know the computer came back with a positive test. That means you can only be in one of these two groups. You don't know which one you are in, but if you were to pretend each laptop were a marble, and were to select one blindly from a bag, you'd have a 50% shot of being in either group because they're both the same size.

Let's plug the pieces into Bayes' theorem and see if the math matches our (new) intuition. We start with Bayes but use the events V and $+$ instead of A and B: $P(V \mid +) \times P(+) = P(+ \mid V) \times P(V)$. Next, we fill in the probabilities we know:

$$P(+) = \text{probability of a positive test} = 198 \text{ positive tests} / 10000 = 1.98\%$$

$$P(+ \mid V) = 99 / 100 = 99\%$$

$$P(V) = 100 / 10000 = 1\%$$

Doing some rearranging of $P(V \mid +) \times P(+) = P(+ \mid V) \times P(V)$, we have:

$$P(V \mid +) \times 1.98\% = 99\% \times 1\% =$$

$$P(V \mid +) = \left(99\% \times 1\%\right) / 1.98\%$$

$$= 50\%$$

That's a lot of math, but we've arrived at the answer: the probability your laptop has the virus given the positive test is 50%.

ENSURE THE PROBABILITIES HAVE MEANING

We've immersed you with numbers and notation through this chapter, especially in the previous section. But now let's take another step back and discuss how to think about and how to use probabilities.

Calibration

When probabilities are defined, they should have meaning.

For instance, assuming equal costs and benefits, a project with a 60% chance of success carries more risk than a project with a 75% chance.

We know this seems obvious, but people often take probabilities like 60% and 75% and mentally cast them as *highly probable* because they are more than 50%. But if this were the case, the probabilities would not be meaningful—they would have been reduced to binary decisions, either an event happened or not, which completely bypasses the point of thinking statistically and dealing with uncertainty.

Moreover, if the probability of an event is 75%, it should happen about 75% of the time.[12] Again, this seems obvious, but it gives your probability meaning. It's a concept called *calibration*. "Calibration measures whether, over the long run, events occur about as often as you say they're going to occur."[13]

Poor calibrations make it impossible to accurately assess risk. If you're a hot-shot lawyer who thinks you're going to win a case with a 90% probability, but historically has only won 60% of cases, you are poorly calibrated and overestimating your chance of success.

So, we say again, probabilities should have meaning. Realize rare events are not impossible and highly probable events are not guarantees.

Rare Events Can, and Do, Happen

A rare event might not happen to you or anyone you know, but it doesn't mean it won't happen at all. Still, we're not always well adjusted to understand events that happen infrequently.

It's true: you are unlikely to win that large jackpot lottery, but the reality is that people do win. When you consider the number of lotteries held around the world every single day, the chances that such a rare event might occur to someone on this earth is not so unrealistic, even if you're not the lucky one.

We often forget the sheer amount of people who live on this planet. With our population on the order of billions, "1 in a million" events seem far more likely to happen. Such events, in fact, capture a lot more people than we can easily comprehend. In a world with 7.8 billion people, a "1 in a million" daily event would happen to 7,800 people per day.

---

[12] We say "about" because there's variation in all things. But in the long run, a 75% event should occur 75% of the time.

[13] fivethirtyeight.com/features/when-we-say-70-percent-it-really-means-70-percent

On the flipside, it's very easy to qualify an activity to the point that it feels rare if only (and perhaps misleadingly) to add a sense of drama. American football, for instance, is filled with heavily qualified commentary suggesting the rarity of an event unfolding on the screen. "This is the first time a 28-year-old rookie has run for 30 yards, after two away games and playing only once in the pre-season." If you put it like that, then yes, it does perhaps seem like an infrequent occurrence.

Do Not Needlessly Multiply Probabilities

Don't needlessly multiply probabilities of past events. You can make anything seem overly improbable.

Let's quickly estimate the probability you are reading this exact line on this page of this book. You're looking at this line out of approximately 35 lines on a page (1/35), on this page out of 300 pages in the book (1/300), and you're reading our book out of *millions*. Multiply those together and you'll get an infinitesimal number. It's clear we were destined for each other!

CHAPTER SUMMARY

This chapter was not only a quick lesson in probability but also a lesson in humility. Probability is hard. But a big part in learning a new topic is respecting that things can go wrong. The information you've learned here will help you seek out additional information before making decisions about probability, especially with decisions that seem intuitive at first but now, we hope, you're suspect of.

In this chapter, we demonstrated how easy it is to misunderstand probabilities. This misunderstanding sometimes comes down to how we word a question—or the assumptions we have behind the information being given. To avoid misunderstandings, remember our guidance when seeing probabilities:

- Be careful assuming independence
- Know that all probabilities are conditional
- Ensure the probabilities have meaning

CHAPTER 7

Challenge the Statistics

Kent Brockman: Mr. Simpson, how do you respond to the charge that petty vandalism such as graffiti is down 80% while heavy sack beatings are up a shocking 900%?

Homer: Oh, people can come up with statistics to prove anything, Kent. Forty percent of all people know that.

—*The Simpsons*

Have you ever heard a statistical claim in the news or in the workplace that you wanted to understand, appreciate, or perhaps challenge? That's what this chapter is about. We're going to teach you about statistical inference, how to use inferential statistics, how to challenge their results, and we'll give you the questions you need to ask to fully understand the underlying inferences being made.

QUICK LESSONS ON INFERENCE

Recall from Chapter 3, "Prepare to Think Statistically," that statistical inference enables us to sample data from the world we live in to then make informed guesses about that world.

In this section, we'll walk you through a series of short examples to show how intuitive statistical inference can be, while gradually adding in formal statistical language (some of which you learned earlier in the book, but a reminder never hurts). The good news is that, no matter your background with statistics, you'll be able to follow the logic of statistical inference we present here.

Give Yourself Some Wiggle Room

Polling data is a common and important example of inferential statistics. You can't survey everyone—you can only poll those in the *sample* of people you have access to. This sample is what we use to help us understand more about the world we live in. For short, we'll say the *sample* helps us understand the *population*.

Let's look at a poll. In Statistics 101 courses across the country, a random sample of 1,000 students are asked, *"Are you sick of statisticians using polling examples to explain basic statistical concepts?"*

The results of our poll are in: 655 students said "yes." (How would you vote?)

Would you feel comfortable proclaiming, based on this one sample of 1,000 students, that the true percentage of all Statistics 101 students (the population) who are sick of polling examples is exactly 65.5? Or do you want some wiggle room in your guess?

You probably wanted some wiggle room. That's good because a week later another 1,000 students are sampled, and this time 670 respond "yes." Sure, 655 and 670 are close, and perhaps you surmise that running these polls has put you close to where the true "yes" percentage of all students might be. The fact is, if you ran this poll many more times, you would get different answers. That's sampling variation. And there's really nothing you can do, except present your results in context. Polling agencies recognize this and place a "margin of error" around the polling results, about +/− 3%, which attempts to capture this fuzziness caused by variation and random chance.

In our first poll, 65.5% is the *point estimate*—and we could present the results as 65.5% +/− 3%, or (62.5%, 68.5%). The interval (62.5%, 68.5%) is called a *confidence interval* and is an example of an inferential statistic. It attempts to use the little information we have in the sample to say something about the big world we live in. We hope the confidence interval captures the true percentage of all Statistics 101 students who are tired of polls.

The lesson: sampling causes variation, which causes uncertainty in your estimate about how many Statistics 101 students are tired of polling examples. Thankfully, confidence intervals give you a range of plausible values for where the true percentage may lie—they give you some wiggle room.

More Data, More Evidence

If you're shopping online and see an Amazon product with a 1-star rating based on a single review, you might ignore the review. It's only one person's opinion. However, if you see a product with a low star rating based on hundreds of reviews, say 300, your view changes. A consensus has started to emerge—the product stinks. So, you choose another product—one with a 4.9-star rating and 200 reviews.[1]

This shows you're already accustomed to understanding how the number of data points behind a statistic as simple as an Amazon star-review affects your trust in it. When referring to the size of the sample (the *sample size*), we'll use the notation N. You didn't trust an N = 1 review but were persuaded by sample sizes of N = 300 and N = 200. As you might imagine, sample size plays an outsized role in statistical inference. Indeed, it would be improbable, though not impossible, for such a product with a 4.9-star rating with N = 200 to be a total piece of junk. But the product with N = 1? That review could be from a random Internet troll.

The lesson: sample size matters. More data means more evidence. (We told you this was intuitive).

Challenge the Status Quo

Fundamentally, science and the creation of new knowledge is about challenging the status quo. When enough evidence accumulates to suggest an old way of thinking is wrong, we adapt. The same is true for statistical inference.

The simplest analogy is the American judicial system. Defendants are "innocent until proven guilty" (the status quo), and only when the evidence suggests, beyond a reasonable doubt, that the status quo is wrong, the defendant is declared "guilty." There's a burden of proof on the prosecution to show the original assumption of innocence is likely wrong.

Researchers, scientists, and businesses use this logic to create new knowledge to better society or their business. Here's how it works. First, they start with a question,[2] like those listed in Table 7.1, and turn it into what's called a *hypothesis test* in statistical inference.

The status quo is called the *null hypothesis*, commonly written as H_0, and this is generally chosen with the hope of throwing it out in favor of new knowledge called the *alternative hypothesis*, written as H_a. The null hypothesis and alternative hypothesis, of course, depend on the question being asked. Table 7.1 breaks down some general questions into appropriate hypothesis tests. Researchers want to find evidence to reject the null in order to support the alternative.

Pay close attention to the setup of the hypothesis tests in Table 7.1. Whatever new knowledge you hypothesize and hope to be true, you start from the assumption that it's *not so* (which would be the status quo). If there exists

---

[1] Don't forget to leave us a review on Amazon.

[2] Recall from Chapter 1, a data science project must start with a clear question.

TABLE 7.1 Questions, Null Hypotheses (H_0), and Alternative Hypotheses (H_a)

Question	Null Hypothesis, H_0	Alternative Hypothesis, H_a
Has MegaCorp's customer satisfaction rate changed in the last quarter?	MegaCorp's customer satisfaction rate *has not changed* in the last quarter.	MegaCorp's customer satisfaction rate *has changed* in the last quarter.
Was MegaBank's SuperBowl commercial effective at increasing year-over-year profit?	MegaBank's SuperBowl commercial *did not change* year-over-year profits.	MegaBank's SuperBowl commercial *changed* year-over-year profits.
Does an experimental COVID-19 vaccine protect against the virus?	The experimental vaccine *is no better than* a placebo.	The experimental vaccine *is better than* a placebo.
Did the unemployment rate in the U.S. change since last month?	The unemployment rate in the U.S. *did not change* since last month.	The unemployment rate in the U.S. *did change* since last month.

enough evidence to show the null hypothesis seems very unlikely, you would reject the null (H_0) and favor the alternative (H_a).

The lesson: hypothesis testing is the hallmark of scientific experimentation. To challenge the status quo, assume it to be true in a null hypothesis. If enough evidence (data) shows the null assumption is unlikely, reject it in favor of new knowledge in the alternative hypothesis.

Evidence to the Contrary

Suppose you're playing a pick-up basketball game with coworkers, and an intern asks to be on your team. He tells you he makes at least 50% of his shots. *Awesome*, you think. Your team needs a good shooter.[3]

Before the game, you make a mental note (i.e., a null hypothesis): the intern's shooting percentage is $\geq$ 50%.

---

[3] We realize 50% is an excellent shooting percentage in basketball. LeBron James, for example, has a career field goal percentage of 50%. So, no, your intern probably can't shoot that well, but 50% makes the math easier to follow. But good for you, being a Data Head and thinking, "Wait isn't that overly optimistic?"

The game starts and you pass him the ball for a wide-open shot. *It's a miss.* No big deal, you think. But then he misses the next shot. Then another. And. . .another. Four misses in a row. *Wow. Just terrible.*

Your trust in him wavers. Is this guy really a baller or is he just pulling my leg? Still, even the professionals have bad days and miss four shots in a row sometimes. So, you keep giving him chances. But the misses pile up. By the end of the game, the intern has missed 10 shots in a row and your team lost. Frustrated, you think, *this guy is a liar.*

You return to your desk and decide to quantify the miserable performance you just witnessed.

So, what are the chances someone who makes 50% of their shots goes on to miss 10 in a row?

Using basic probability, you crunch some numbers. The chance he misses one shot is 50%. The chance he misses two in a row is 50% × 50% = 25% (assuming each shot is independent, using the rules from the previous chapter). Continuing this pattern, you multiply 50% by itself 10 times, or $0.5^{10} = 0.00098$, so 0.1%, or approximately 1 in 1,000.

Therefore, the probability of seeing this specific result—10 missed shots—*given* that he can supposedly make 50% of his shots, is 1 in 1,000. This probability, 1 in 1,000 or 0.001, is called a *p*-value (*p* for probability). Now you must decide, do you think the intern had an unlucky day? Or was your null hypothesis (the intern's shooting percentage is ≥ 50%) just wrong?

Ten missed shots just strains credulity. A 1-in-1,000 kind of bad luck day is strong enough evidence that the original claim is unlikely to be so. In fact, you probably rejected the null hypothesis earlier in the game and accepted the alternative hypothesis, H_a: The intern's shooting percentage is < 50%.

Take a moment and ask yourself: when did you start to second guess your intern instead of making excuses for him? What was your cutoff for the missed number of shots to reject the null hypothesis?

For the sake of example, let's say your cutoff was at 5 missed shots. If the intern had only missed 4 in a row, which has probability $50\%^4 = 6.25\%$, or 1-in-16, you could have given him the benefit of the doubt. But once he missed 5 shots, there was too much evidence to the contrary he was a good shooter. That cutoff, 5 missed shots in a row, surpassed what is called the *significance level*. The data was no longer consistent with the claim.

Because the universe is filled with variation, you must accept some level of randomness (and missed shots). Sometimes someone could just play poorly without a clear explanation as to why. In that way, a significance level is an

---

[4] O'Neil, C., & Schutt, R. (2013). *Doing data science: Straight talk from the frontline.* O'Reilly Media, Inc.

artificial threshold, decided upon by you, in which you can tolerate randomness and unexplained variation but still feel the null hypothesis is true. If the p-value is less than the significance level, you reject the null and say the result is *statistically significant.*

The lesson: testing whether a p-value was less than a significance level to reject a null hypothesis is a key part of statistical inference. Of course, the presence of variation and choosing a significance level opens you up to potential decision errors.

Balance Decision Errors

When variation leads to the wrong conclusion, this is called a *decision error.*

There are two types of decision errors, both given the nondescript names: Type I (*false positive*) and Type II (*false negative*) errors. Because being descriptive is important, we prefer false positive and false negative in place of Type I and Type II errors.

What is a false positive error? It's when evidence appears to confirm the reality of the alternate hypothesis when instead it should have been rejected (e.g., a man has a positive pregnancy test). On the other hand, a false negative error happens when you accept a false null (e.g., a pregnant woman has a negative pregnancy test). Table 7.2 provides more examples of both types of errors.

You, as a decision maker, choose the probability of false positives by setting the significance level. Related to statistical significance is something called *power*, the probability of correctly rejecting the null hypothesis when

TABLE 7.2 False Positive vs. False Negative Decision Errors

Question	Null Hypothesis	False Positive	False Negative
Did the defendant commit a crime?	Defendant is innocent.	Sending an innocent person to jail.	Letting a guilty defendant go free.
Do you have a disease?	You do not have a disease.	You test positive even though you do not have the disease.	You have the disease, but the test did not detect it.
Has MegaCorp's customer satisfaction rate changed in the last quarter?	This quarter's recommendation rate $\leq$ Last quarter's recommendation rate.	This quarter's results showed an improvement by chance alone.	This quarter's results did improve, but the test did not detect it.

the alternative hypothesis is true. The higher the power of a test, the lower the probability of a false negative.

False positive and false negative errors have inherent tradeoffs, and—unless you collect more data—you cannot decrease one without increasing the other. For example, you probably want a low false positive rate in your spam. The null hypothesis here is "the email is not spam," so a false positive would mean an email from your mother is in the spam folder. This comes with the price of more spam emails in your inbox (more false negatives), but you can tolerate that if you get most of your personal email. In disease screening, however, the medical community may decide to accept more false positives in order to decrease the false negatives (a missed diagnosis). If someone has a disease, they want to detect it.

The lesson: variation complicates decision making. You will sometimes think your alternative hypothesis is true when it's not (false positive) or think the null hypothesis is true when it's not (false negative).

THE PROCESS OF STATISTICAL INFERENCE

So far, we've discussed many of the components of statistical inference scattered throughout five quick lessons. You may be wondering how all these components fit together. Let's see if we can summarize them so that you, as a Data Head, can clearly understand and communicate the process of statistical inference.

In short, statistical inference follows these steps:

1. Ask a meaningful question.
2. Formulate a hypothesis test, setting the status quo as the null hypothesis, and what you hope to be true as the alternative hypothesis.
3. Establish a significance level. (5% or 0.05 is an arbitrary but often-used number.)
4. Calculate a p-value based on a statistical test.
5. Calculate relevant confidence intervals.
6. Reject the null hypothesis and accept the alternative hypothesis if the p-value is less than the significance level; otherwise, fail to reject the null.

Pause for a quick reflection. If you can read and understand those six steps, congratulations! You are learning the language of statistics. The one part we casually glossed over is the idea of a *statistical test*. These are mechanisms that calculate the p-values. We did it with basic probability in the intern basketball example (taking 50% to the 10th power). But there are hundreds of statistical tests used to describe, compare, assess risks, and assess relationships in your data. These are the tools in the toolbox that most Statistics textbooks focus on. We did not focus on statistical tests here because you can and should understand the logic behind statistics regardless of how the *p*-values are calculated.

Returning to the task at hand, we recognize that Data Heads will often be the consumers of statistical results rather than the creators of them. So in the next section, we're going to teach you the questions you should ask to challenge the statistics you see. If you followed along in the previous sections, you should feel empowered and prepared to ask these questions.

THE QUESTIONS YOU SHOULD ASK TO CHALLENGE THE STATISTICS

We created the following list of questions and prompts to ask your teammates to help you confidently challenge the statistics presented to you:

- What is the context for these statistics?
- What is the sample size?
- What are you testing?
- What is the null hypothesis?
- What is the significance level?
- How many tests are you doing?
- Can I see the confidence intervals?
- Is this practically significant?
- Are you assuming causality?

Let's go through each of these questions and why they're important.

What Is the Context for These Statistics?

The context surrounding statistics are just as important as the numbers themselves. If you hear, "Sales are up 10%!" your brain should reflexively think, "Compared to what?"

Consider this example: A marketing analyst tells his boss sales are up 10% compared to last quarter but fails to share that sales from its largest competitor grew 15%. Surely the boss wants the additional context. But efforts to summarize can obfuscate. Data Heads should demand context and baselines to compare against.

Let's look at another example. Suppose a new YouTube commercial increases the probability someone clicks on an advertisement by 50%. Without context, this sounds quite impressive. Putting the statistic into context, however, reveals that the commercial click-through rate (how many people clicked on the ad divided by how many people saw the commercial) improved from 0.1% to 0.15% (10 in 10,000 compared to 15 in 10,000), an absolute increase of 0.05%. *It should be reported this way.* Reporting the relative percentage change, (0.0015 − 0.0001)/0.0001 × 100 = 50%, gives the wrong impression.

You've probably come across examples like this in your own line of work, where you see a statistic—precise, definitive, and impressive—but you don't know what it really means. Don't be afraid to speak up and ask, "What is the context for these statistics?"

What Is the Sample Size?

By now, you understand why sample size is important. When N is small, you usually see a lot of variation. No problem—you'll just add more data to the pile. Enough data that the results will have less variation, right? In the age of "big data," you might be tempted to just make N so enormous it could cover every possibility.

Whenever N is large, it's easy to make the leap and think N = ALL; every possible data point is at your disposal. But believing N=ALL does not absolve you from thinking about data quality and bias. (Remember the lessons in Chapter 4, "Argue with the Data.") Are you truly capturing people from the population you care about?

As noted in the book *Doing Data Science: Straight Talk from the Frontlines*:[4]

> Indeed, we'd argue that the assumption we make that N=ALL is one of the biggest problems we face in the age of Big Data. It is, above all, a way of excluding the voices of people who don't have the time, energy, or access to cast their vote in all sorts of informal, possibly unannounced, elections.

Excluding the voices doesn't only pertain to elections. Those in need may be incorrectly excluded from receiving discounts on food or clothes; being included in surveys about public policy; or simply not being counted. It's easy to assume the large enough dataset is a true reflection of a population, but size isn't everything. Worse, having "big data" makes it easy—too easy, in fact—to find spurious relationships. Slice and dice the data in enough ways, and surely something interesting will pop out.

In the rare cases when N truly equals ALL the population (aka a census), then you are in luck. You don't need statistical inference because there would not be uncertainty in the descriptive statistics, assuming the data was collected correctly.

What Are You Testing?

Underlying any statistical inference claim in the workplace or in the news is (*we hope*) a specific question that can be tested with data. Don't let a data worker share numbers without sharing the underlying question. Make sure the team knows *why* statistics are being generated in the first place. Ask "What are you testing?" and expect an answer in clear, non-statistical terms.[5]

What Is the Null Hypothesis?

Your intern at MegaCorp has been working closely with the customer service department this quarter, offering ideas to boost customer satisfaction. You want to see if his ideas are improving MegaCorp's customer satisfaction, which is measured via a simple, one-question survey: "Would you recommend us to a friend?"

The intern formalizes the test and establishes a null hypothesis: "This quarter's recommendation rate is no worse than last quarter's rate." Thus:

- H_0: This quarter's recommendation rate $\geq$ Last quarter's recommendation rate.

If the null hypothesis is rejected, then the *alternative hypothesis* will be accepted. In this case, the alternative hypothesis is "This quarter's recommendation rate is worse than last quarter's rate." Using statistical notation, the alternative hypothesis, denoted H_a, would be written as:

- H_a: This quarter's recommendation rate $<$ Last quarter's recommendation rate.

Pause here and think about what's been assumed. You have not seen any data or statistics, but you're able to challenge the logic of your intern's approach. In establishing the null hypothesis, he has set himself up for victory by default. If the survey results between the two quarters are relatively close or are based on a small number of customers (i.e., a small sample size), then there may not be enough evidence to reject the original assumption. This mistake is why Data Heads need to ask, "What is the null hypothesis?" A poorly defined null can create the deception where something is right simply because no data exists to show it's wrong.

Remember, science is about challenging the status quo. In practice, the null and alternative hypotheses should be established to have the alternative

---

[5] Review Chapter 1 for more on how to interrogate the question itself.

hypothesis reflect what you might believe or hope to be true. This puts the burden of proof on the data to show the null hypothesis is unlikely.

Your intern, wanting to show his efforts to improve customer service made a measurable difference, should use the following setup for his hypothesis test:

- H_0: This quarter's recommendation rate $\leq$ Last quarter's recommendation rate.
- H_a: This quarter's recommendation rate $>$ Last quarter's recommendation rate.

(We'll return to this example shortly.)

Assuming Equivalence

Suppose you replace the key ingredient in a food product to save costs. Your team conducts a quick taste survey, based on a 10-point scale, to see if customers notice. With the previous recipe, 18 out of 20 people said they'd buy the product. In a new survey with the new recipe, 12 out of 20 said they would buy it.

Under the null hypothesis: "New product's buy rate = previous product's buy rate" and a significant level of 0.05, a statistical test calculates a p-value[6] of 0.064. The p-value is above 0.05, so the null is not rejected. Your boss George takes this as, "My data team showed there's no statistical difference between our old recipe and this new, cheaper recipe. Let's cut costs."

George is assuming equivalence between the old recipe and new recipe, but he may not have enough data yet to show they are different. The lesson here: failing to reject the status quo is not the same as proving it.[7]

What Is the Significance Level?

Recall the significance level as the threshold at which we can tolerate the data as being inconsistent with the null hypothesis.

The conventional, and somewhat arbitrary, significance level historically has been 5%, or 0.05. Other industries or researchers may use 1%, or 0.01. Some go even smaller. At the European Organization for Nuclear Research (CERN), researchers used an incredibly small significance level when testing

---

[6] We used a two-sided Fisher's exact test.

[7] This example would require something called an equivalence test. It's beyond the scope of this chapter but be aware it's out there. Tell your team and use it. If you can follow the logic of this chapter, you can understand an equivalence test.

for the tiny physics particle known as the Higgs boson.[8] The smaller the significance level, the smaller the chance of a false positive.

Chances are, you'll start at the 5% significance level because you have to start somewhere. However, the choice matters. A 5% significance means you can tolerate that you will wrongly reject a null hypothesis (have a false positive) 1 out of 20 times. Does that make you feel comfortable?

Perhaps it's all too easy to pick a significance level that ensures your results are always statistically significant. Many tools already have a 5% level programmed in as a default. That level may not even be reflective of your industry—or, the level might have been set by your data scientist who did not communicate the change to you—only that the result was statistically significant. In the worst case, someone may run a test and select a significance level after the fact—this is like throwing a dart and then moving the target over it. For example, someone might run a statistical test, get a *p*-value of 0.11, and then set their significance level at 0.15 to get a statistically significant result.

Which is why it's always important to ask, "What is the significance level?"

In a more practical sense, know that lowering a significance level from, say, 5% to 1%, will decrease your false positives. That means you've set a higher bar to reject the null hypothesis. The data then must be more extreme (or, at the very least, convincing) to reject the null hypothesis. Doesn't sound so bad, right? Well, it comes at the cost of increasing your false negatives. The tradeoff is not one to take lightly, and there is no one-size-fits-all recommendation. The proper balance depends on your problem and your ability to absorb the impacts of false negative and false positive errors.

How Many Tests Are You Doing?

After knowing the significance level, you also want to ask how many tests your data workers are doing. As they look at the data in different ways, they may end up running dozens, maybe hundreds of informal statistical tests at the 5% significance level. Suppose, for example, a researcher is testing a large dataset of cancer patients and the types of food they eat, searching for which foods might be associated with greater survival rates. With 100 different types of food in the database and a 5% significance level, 5 foods would appear as statistically significant in the fight against cancer *even if no foods have a true effect.*[9]

---

[8] "5 Sigma What's That?" blogs.scientificamerican.com/observations/five-sigmawhats-that

[9] There are statistical ways to correct this. Look up the "Multiple Comparisons Problem."

Can I See the Confidence Intervals?

We've talked a little about confidence intervals and some of their components. Now we'll put the pieces together.

What do we mean by the word *confidence*? This, like the word *significance*, has a different meaning in statistics than it does in everyday use. In statistics, significance and confidence are intrinsically linked. In fact, there's a symmetry between the *significance level* and *confidence level*—a significance level of 5% corresponds to a confidence level of 95%. More formally, the *confidence level* = 1 − the *significance level*. Instead of hearing, "We rejected the null hypothesis using a 5% significance level," you might hear, "We rejected the null with 95% confidence."

Next, let's understand why someone looking at statistical results should ask to see the confidence intervals. A confidence interval, as you may recall, tries to capture the true number you're after in a population. The 95% confidence interval in the polling example earlier in the chapter with sample size N = 1,000 was (62.5%, 68.5%). Suppose instead of sampling 1,000 students, we only had access to 100 in our poll, and 65% said yes. The resulting 95% confidence interval is (54.8%, 74.2%). The confidence interval is much wider than the original confidence interval with 1,000 people because we have a smaller sample size. This means we need to cast a wider net around where we think the percentage in the population may lie. But if the sample size N increases, the confidence interval would tighten. More data means more evidence and less uncertainty. Makes sense, right? If the entire population responds, a confidence interval isn't needed—you've found the true population number.

Confidence intervals also provide estimate of *effect sizes* in a statistical test.[10] Suppose you want to know if women basketball players from the United States are the same height as women players from Europe. You set up your null hypothesis and alternative hypothesis:

- H_0: mean height of U.S. players = mean height of European players
- H_a: mean height of U.S. players ≠ mean height of European players

Now imagine your data worker goes off, collects data, and calculates a *p*-value to compare against a 5% significance level. They come back with the results: the *p*-value was less than the significance level. U.S. players and

---

[10] Effect size can take on many meanings in statistics. Here, we're simply talking about effect size as the difference between two numbers.

European players are not the same height, and the results are statistically significant.[11]

But don't you feel like you're missing some information? Sometimes we view statistical significance as the seal of approval. *Oh, your results are statistically significant? That means they're 100% true.* But statistical tests are looking for any difference, whether important or not. This is why you should never be satisfied with just *p*-values. In the basketball example, suppose the mean heights for U.S and European players are 72 inches and 71.5 inches, respectively, and the 95% confidence interval around their differences is 0.5 +/− 0.4 inches.

Is the effect size, a mere half-an-inch, a practical or interesting difference?

Is This Practically Significant?

Trivially small effects can be found with large sample sizes. If you are only seeing *p*-values and not confidence intervals, you might think you've found a large effect, when what you really detected was a minor difference that has no practical value. So as you're looking at the confidence intervals, ask if what you're seeing is a meaningful, practical effect.

Are You Assuming Causality?

You almost forgot about the intern. You're curious if his work with customer service has improved the recommendation rate of your customers from last quarter to this quarter. Because you'd like to see some evidence for improvement, the intern has established the null and alternative hypothesis as:

- H_0: This quarter's recommendation rate $\leq$ Last quarter's recommendation rate.
- H_a: This quarter's recommendation rate $>$ Last quarter's recommendation rate.

Each quarter's survey had a sample size of 100. 50/100 customers would recommend the company last quarter, and 65/100 would recommend the company this quarter. Are the results statistically significant at the 5% level?

Using a statistical test,[12] the intern calculates a *p*-value of 0.02, less than 0.05, so you can reject the null and accept that this quarter's results are statistically different than last quarter. The intern is excited and feels he's avenged

---

[11] No, we did not actually run this test or collect data.

[12] Using the statistical software R: `prop.test(c(65, 50), c(100, 100), alternative = "greater")`

his poor performance on the basketball court. "So it looks like my customer service intervention really worked."

But has it? Correlation is not causation. Customer satisfaction may have been improved by a number of factors, and unless there was a designed experiment performed and great care to measure the difference of the old approach against the intern's ideas, then you do not have causality.

CHAPTER SUMMARY

In this chapter, you learned about statistical inference and how to challenge the statistics that come your way. Specifically, you learned what questions to ask about statistical claims and why it's important to ask them. The questions you should ask are:

- What is the context for these statistics?
- What is the sample size?
- What are you testing?
- What is the null hypothesis?
- What is the significance level?
- How many tests are you doing?
- Can I see the confidence intervals?
- Is this practically significant?
- Are you assuming causality?

Armed with this list, you'll be able to effectively challenge, understand, and appreciate the statistics you see.

Understanding the Data Scientist's Toolbox

You know the terms: machine learning, artificial intelligence, and deep learning. They're probably what led you to this book. But now we're going to demystify them.

Our data field, whatever name we choose to call it, is in a constant state of change. But there are fundamental concepts and tools that have persisted for decades and form the foundation for today's hottest trends, including text and image analysis. Part III, "Understanding the Data Scientist's Toolbox," teaches you those concepts and techniques at a high level.

Here's what we'll cover:

Chapter 8: *Search for Hidden Groups*

Chapter 9: *Understand the Regression Model*

Chapter 10: *Understand the Classification Model*

Chapter 11: *Understand Text Analytics*

Chapter 12: *Conceptualize Deep Learning*

You'll also learn about common mistakes and traps that even seasoned analysts fall into.

III

Understanding the Data Scientist's Toolbox

You know the terms machine learning, artificial intelligence, and deep learning. They're probably what led you to this book. But now we're going to demystify them.

Our data field, whatever name we choose to call it, is in a constant state of change. But there are fundamental concepts and tools that have persisted for decades and form the foundation for today's hottest trends, including text and image analysis. Part III, "Understanding the Data Scientist's Toolbox," introduces you these concepts and techniques at a high level.

Here's what we'll cover:

You'll also learn about common mistakes and traps that even data science analysis fall into.

CHAPTER 8

Search for Hidden Groups

"If you mine the data hard enough, you can find messages from God."

—*Dilbert*[1]

Imagine you get a call from a friend. They're looking for help categorizing their music collection—a vintage set of vinyl records. You agree to help.

As you drive to your friend's house, you wonder how you would organize such a collection. You could start with some obvious categories. For instance, music is often organized into genres and subgenres. Or, you could group them by the musical periods in which they came out. This information is readily available on an album cover.

When you arrive at your friend's house, however, you are handed a tall stack of black vinyl records—no album covers to be found.

You learn your friend purchased the stack of records at a yard sale and has no clue which (or how many) genres, artists, or musical periods it contains. Now you must leave your preconceived notions about how to categorize the records at the door—you no longer have predefined groups on the album cover to guide you. The task of categorizing records is suddenly much more difficult than you anticipated.

Determined, you and your friend break out the record player, listen to each album, and start grouping them into categories based on how similar

---

[1] Adams, Scott. Dilbert Cartoon. January 03, 2000.

they sound. As you listen to the records, new groups emerge, small groups might combine into one, and occasionally, a record moves from one group to another after a spirited debate about which group it sounds "closest" to.

In the end, the two of you settle on 10 categories and give each a descriptive name.

You and your friend just performed what's known as *unsupervised learning*. You didn't come in with preconceived notions about the data, but instead let the data organize itself.[2]

This chapter is all about unsupervised learning: a collection of tools designed to discover hidden patterns and groups in datasets when no predefined groups are available. It's a powerful technique used in a variety of fields, from segmenting customers into different marketing groups, to organizing music on Spotify or Pandora, and organizing the photos on your phone.

UNSUPERVISED LEARNING

At the core of unsupervised learning is the idea that hidden groups are lurking beneath data. There are a lot of ways to twist, turn, and replot this data to identify these interesting patterns and groups—*if groups exist*. As a Data Head, you must be able to navigate across the many unsupervised learning methods in your search for hidden groups in data.

But how does one even start? The vast amount of unsupervised learning techniques available seems daunting. Luckily for you, you just need a basic understanding of the main activities related to unsupervised learning. So, here we draw on two activities and a foundational technique for each:

- Dimensionality reduction through principal component analysis (PCA)
- Clustering through *k*-means clustering

In this chapter, we'll go through these techniques, what they mean, and how they achieve their goals of dimensionality reduction and clustering, respectively.

DIMENSIONALITY REDUCTION

Dimensionality reduction is a process you're already familiar with. Photography is an example; it reduces the three-dimensional world down to a flat, two-dimensional photo you can carry in your pocket.

---

[2] *Kind of.* It's not quite that easy.

With datasets, we're working with rows and columns: observations and features. The number of columns (features) in a dataset is called the *dimension* of the data, and the process of condensing many features into new and fewer categories, while retaining information about the dataset, is called *dimensionality reduction*. Simply speaking, we're looking for hidden groups in the columns of a dataset so that we can combine multiple columns into one.

Let's talk about why this matters. Practically speaking, it's hard to make sense of datasets with many features. They can be slow to load on a computer and a pain to work with. Exploratory data analysis becomes tedious, and in some cases, infeasible. In bioinformatics, for example, the potential dimension of a dataset can be enormous. Researchers may have thousands of gene expressions for each observation, many of which are highly correlated (and thus possibly redundant) with each other.

The desire to speed up computational time, remove redundancies, and improve visualization tells us why dimensionality reduction of data matters. But how can we do it?

Creating Composite Features

One way to reduce the dimension of a dataset is to combine multiple columns into a composite feature. Let's do this by exploring some real-world data on automobiles from a 1974 Motor Trend road test of 32 automobiles and 11 features like miles-per-gallon, horsepower, weight, and other vehicle attributes.[3] Our task is to create an "efficiency" metric to rank the cars from most-efficient to least-efficient.

Miles-per-gallon (MPG) seems like an obvious place to start the exploration. This is plotted in the leftmost chart of Figure 8.1. If you look at the distribution in that first chart, you can see some separation with the best MPG car on top and the worst on the bottom, but you also notice many vehicles tend to group around the center. Can we layer in additional information to further separate the data? Let's move to the middle chart. Here we've created a composite feature: a car's MPG minus its weight,[4] MPG – Weight. Notice there's much more spread simply by grouping two features into one composite.

Next, let's go a step further and create a third efficiency metric, *Efficiency = MPG – (Weight + Horsepower)*. (An equation like this is called a *linear*

---

[3] This is the mtcars dataset from the R software. stat.ethz.ch/R-manual/R-devel/library/datasets/html/mtcars.html. For visualization purposes, we are only displaying 15 cars, not all 32.

[4] Because the features have widely different ranges, they need to be put on a similar numeric scale before being combined.

Visualizing Different Data Components for Cars

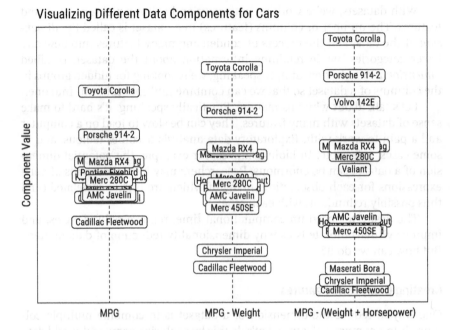

FIGURE 8.1 Sorting cars based on different composite features. Notice how the cars spread out from each other (aka have higher variance) as more features are condensed into a single "efficiency" dimension.

combination.) This combination of columns has separated the data more so than the other features. It gives us more information—more spread—about cars and has uncovered something interesting. You can see the heavy, gas-guzzlers at the bottom and the light, fuel efficient cars on top. We've effectively created a new dimension of the data (efficiency) by combining the original dimensions, and this new dimension lets us ignore the three original dimensions. That's dimensionality reduction.

In this example, we had the foresight to know that combining a car's MPG, Weight, and Horsepower into a new composite variable would begin to uncover something interesting in the data. But what if you don't have the luxury of knowing which features to combine or how to combine them? That's indeed the nature of unsupervised learning, and that's where principal component analysis enters the picture.

PRINCIPAL COMPONENT ANALYSIS

Principal component analysis (PCA) is a dimensionality reduction method invented in 1901,[5] long before the terms *data scientist* and *machine learning* became part of business terminology. It remains a popular—though often misunderstood—technique. We'll attempt to clear up this confusion and set the record straight about what it does and why it's useful.

Unlike our approach in the car example, the PCA algorithm does not know in advance which groups of features to combine into composite features, so it considers all possibilities. Through some clever math, it layers the dimensions in different configurations, looking for which linear combinations of features spreads the data out the most. The best of these composite features are called the *principal components*. Better yet, the principal components represent new dimensions in the data that are not correlated with each other. If we ran PCA on the cars data, we might not only uncover an "efficiency" dimension, but also a "performance" dimension.

You might wonder how you can tell if PCA can combine a dataset's features into meaningful groups that create principal components. What exactly is PCA looking for?

Let's consider these questions in the next thought experiment. And because your authors know little to nothing about cars, we're going to teach the next lesson with a different (and hypothetical) dataset.

Principal Components in Athletic Ability

Imagine that you work at an athletic performance camp. You have a spreadsheet with hundreds of rows and 30 columns. Each row contains information about an athlete's fitness: the number of push-ups, sit-ups, and deadlifts they can do in one minute; how long it takes them to run 40 meters, 100 meters, 1,600 meters; various vital signs like resting pulse and blood pressure; and several other performance and health metrics. Your boss gave you the vague task to "summarize the data," but you're bogged down by the sheer number of columns in the spreadsheet. No doubt, there's a wealth of information here, but could you reduce these 30 features into a more reasonable number you could use to summarize and visualize this data?

---

[5] Pearson, K. (1901). LIII. On lines and planes of closest fit to systems of points in space. *The London, Edinburgh, and Dublin Philosophical Magazine and Journal of Science*, 2(11), 559–572.

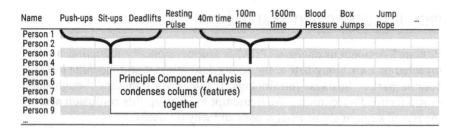

FIGURE 8.2   Principal component analysis groups and condenses the *columns* of a dataset into new, uncorrelated dimensions.

To start, you spot a few obvious patterns. Athletes who can do the most push-ups generally do the most deadlifts, and those who are sluggish in the 100m dash also struggle in the 40m sprint. It looks like many of the features are indeed correlated with each other because they measure related performance properties. You suspect, since several groups of these features are correlated with each other, there might be a way to condense them into fewer dimensions that are no longer correlated with each other but contain as much information as possible from the original data. *This is precisely what PCA does!*

See Figure 8.2 for a high-level overview of what you're trying to accomplish. However, you can't easily explore the correlations of 30 variables, even with a computer. (This would require 435 separate scatterplots to view each pair of features.[6]) So, you run the data through the PCA algorithm to exploit the embedded correlations in the dataset for you. PCA returns two datasets as an output.[7]

Figure 8.3 shows the first dataset. This table has the athletes' features along the rows of the dataset. The columns show *weights* of those features and how they roll into a principal component. These weights reflect an important step in PCA creating new dimensions in the data. (Note, *weights* here refers to a term specifically used in PCA and not lifting weights.) We've taken the liberty of visualizing the weights; however, in numeric terms they are measurements of correlation, with a range of −1 to 1. The closer the numbers are to either extreme, the stronger the correlations, and the more each original feature contributes. So, what you're looking for are interesting patterns in the weights of the principal components (denoted PC in the figure): weights that are far away from the vertical line of zero might tell a story.

---

[6] 30 choose 2 = $30!/((30 − 2)! \, 2!) = 435$

[7] No software will return PCA results as we're showing them here. We are going to great lengths to avoid equations and numbers and therefore have decided to focus on visualizations.

In the first column, "Weights for PC 1," you see high weights for push-ups, sit-ups, deadlifts. These three are positively correlated with each other, as you spotted earlier. PCA automatically detected this, too. You might decide to call this combination of features "Strength" based on its underlying features. As you look at the weights for PC 2, you notice the negative bars are associated with measures of "Speed" (low resting pulse, low 40m times, low 100m times). Similarly, you might name PC 3 "Stamina" and PC 4 "Health."

Before you had multiple dimensions with correlations. However, these four new dimensions represent four composite features that are uncorrelated with one another. And because they are not correlated, that means each new dimension provides *new, nonoverlapping* information. These dimensions effectively partition the information in the dataset into distinct dimensions, as shown in the row "% of Information for Each Component." Using just these 4 new features, we can maintain 91% of the information in the original dataset.

Using the weights from Figure 8.3, each athletes' 30 original measurements can then be transformed into the principal components of "Strength," "Speed," "Stamina," and "Health" using linear combinations. For example, an athlete's strength is calculated by:

Strength = 0.6*(number of push-ups) + 0.5*(number of deadlifts) + 0.4*(number of sit-ups) + (minor contributions from other features)

Again, the numbers (the weights) 0.6, 0.5, and 0.4 are given to you from PCA. (We just chose to visualize these numbers instead.)

Doing this series of calculations for all athletes produces the second output from the PCA algorithm, as shown in Figure 8.4. It's a new dataset, the same size as the original, but as much information as possible has been shifted to the first handful of uncorrelated principal components (aka composite

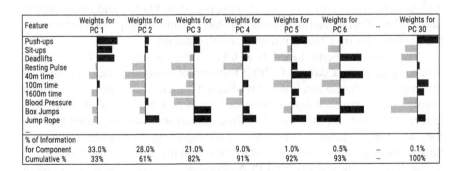

Feature	Weights for PC 1	Weights for PC 2	Weights for PC 3	Weights for PC 4	Weights for PC 5	Weights for PC 6	...	Weights for PC 30
Push-ups								
Sit-ups								
Deadlifts								
Resting Pulse								
40m time								
100m time								
1600m time								
Blood Pressure								
Box Jumps								
Jump Rope								
...								
% of Information for Component	33.0%	28.0%	21.0%	9.0%	1.0%	0.5%	...	0.1%
Cumulative %	33%	61%	82%	91%	92%	93%	...	100%

FIGURE 8.3 PCA finds optimal weights that are used to create composite features that are linear combinations of other features. Sometimes, you can give the new composite feature a meaningful name.

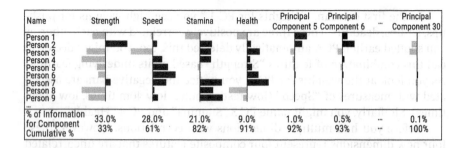

Name	Strength	Speed	Stamina	Health	Principal Component 5	Principal Component 6	...	Principal Component 30
Person 1								
Person 2								
Person 3								
Person 4								
Person 5								
Person 6								
Person 7								
Person 8								
Person 9								
...								
% of Information for Component	33.0%	28.0%	21.0%	9.0%	1.0%	0.5%	...	0.1%
Cumulative %	33%	61%	82%	91%	92%	93%	...	100%

FIGURE 8.4 The PCA algorithm creates a new dataset, the same size as the original, where the columns are composite features called principal components.

features). Notice how the contribution by latter principal components drops off after the fourth component.

Consequently, rather than using 30 variables to explain 100% of the information in the original dataset, the dataset in Figure 8.4 can explain 91% of that information in only 4 features. Therefore, we can decide to safely ignore 26 columns. That's dimensionality reduction! Armed with this dataset, you could sort the four columns to find out who is the strongest, who is the fastest, or any combination therein. Visualizations and interpretation are now much easier.

PCA Summary

Let's zoom back out to clarify a few things.

First, for a column in a dataset, a good proxy for *information* is *variance* (a measure of spread). Think of it this way. Suppose we add a new column to the athlete dataset in Figure 8.2 called "Favorite Shoe Brand" and every person said "Nike." There'd be no variation in that column to help differentiate one athlete from another. No variation = no information.

The core idea behind PCA is to take all the variance—all the information—available in a dataset (a high number of columns) and condense as much of that information into as few distinct dimensions as possible (a lower number of columns). It does this by looking at how each original dimension is correlated with another. Many of these dimensions are correlated because they so strongly measure the same underlying thing. In this sense you only have a few true dimensions of data that maintain most of the information in the dataset. The math behind PCA effectively "rotates" the dimensions in such a way that we can look at it in fewer dimensions (the principal components) without losing a lot of information.

This is not unlike taking a picture. For instance, you could take a picture of the Great Pyramids of Egypt from countless angles, but some angles are more descriptive than others. Take a photo overhead with a drone, the pyramids appear as squares. Take a photo directly in front, they look like triangles. Which rotation of the camera would capture the most information to impress your friends when you condense the 3D world in Giza down to a 2D photo on your smart phone? PCA finds the best rotation.

Potential Traps

Now that you are more familiar with PCA, we confess that, in real-world data, there will never be such a clean separation of groups into easily distinguishable principal components like we shared in the fitness example.

Data is messy, so the resulting principal components will often lack clear meaning and may not be able to have descriptive nicknames. In our experience, people try too hard to find catchy titles for principal components to the point where they often create a picture of the data that doesn't exist. As a Data Head, you should not accept immediate definitions for your principal components. When others present already-named components, challenge their definitions by asking to see the equations behind the groupings

Moreover, PCA is not about knocking out variables that aren't important or interesting. We see people make this mistake often. The principal components are generated from *all* the original features. Nothing has been deleted. In the fitness example, every original feature might be grouped with several others to form the four main principal components: *Strength, Speed, Stamina, Health*. Remember, the resulting dataset from the PCA calculation is the same size as the original dataset. It's up to the analyst to decide when to drop the uninformative components, and there's no right way to do this. This means, if you see results from PCA, ask how they decided how many components to keep.

Finally, PCA relies on the assumption that high variance is a proxy for something interesting or important within the variables. In some cases, it's not a bad assumption. But it isn't always the case. For instance, a feature can have high variance and still have little practical importance. Imagine, for example, adding a feature to the data with the population of each athlete's hometown. This feature, while highly variable, would be completely unrelated to the athletic performance data. Because PCA is looking for the wide variation, it might mistake this feature as being important when it's not.

CLUSTERING

Groups of features (columns) might tell one story, like with PCA, but groups of observations (rows) might tell a different story. This is where clustering comes in.[8]

Clustering, in our experience, is the most intuitive data science activity because the name is truly descriptive of the task (unlike the name principal component analysis). If your boss told you to cluster our friendly camp of athletes into groups, you'd understand the task. Some natural questions would arise when you'd look at the data in Figure 8.5: How many groups might exist? How would you categorize the groups? Yet you would be able to get started. Perhaps the stronger and slower athletes form one group, while the weaker and faster form another. You might name the groups "Body Builders" and "Distance Runners."

Take a minute to think about how you would go about clustering that data and the decisions you'd make along the way. If you were lazy, you could say, "Every person in this sheet is an athlete, so there is just one group: athletes." Or, if you were even lazier, you could say, "Every person is their *own* group. There are N groups." Both are unhelpful. So, at this point, you've established the obvious: you need between 1 and N groups.

Another decision you would have to make, without a clear set of guidelines, is how to tell if one athlete is "similar" or "close" to another. Consider the subset of data in Table 8.1. Which two of these athletes are closest to each other?

You could make an argument for any pairing. It all depends on how you judge "close." Athletes A and B are close in push-ups and pulse. A and C are close because they're each the best at something, 1,600m time and number of push-ups, respectively. And B and C are close because they run slower. Here,

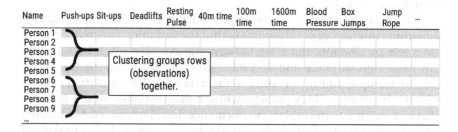

Name	Push-ups	Sit-ups	Deadlifts	Resting Pulse	40m time	100m time	1600m time	Blood Pressure	Box Jumps	Jump Rope	...
Person 1											
Person 2											
Person 3											
Person 4											
Person 5			Clustering groups rows								
Person 6			(observations)								
Person 7			together.								
Person 8											
Person 9											
...											

FIGURE 8.5 Clustering is a technique that groups *rows* of a dataset together. Recall that PCA grouped the *columns* together.

---

[8] For clarity, PCA and clustering are separate objectives. One does not need to be done to use the other.

TABLE 8.1 Which of These Two Athletes are "Closest" to Each Other?

Athlete	Push-Ups	Resting Pulse	1600m Time
A	40	50	4:30min
B	30	55	8:00min
C	100	65	9:00min

you see what you want to see. It depends on which features matter most to you and your background and how you internally measure the concept of "close." Unsupervised learning, of course, knows none of that.

This example demonstrates some important issues in clustering: How many clusters should exist? How can any two observations be considered "close" to each other? And what's the best way to group close observations together?

k-means clustering is a way to start.[9]

K-MEANS CLUSTERING

k-means is a popular clustering technique among data scientists. With *k*-means, you tell the algorithm how many clusters you want in your data (the *k*), and it will group your N rows of data into *k* distinct clusters. The data points within a cluster are "close" to each other, while the distinct clusters are as "far apart" from each other as possible.

Confused? Let's consider an example.

Clustering Retail Locations

A company wants to assign its 200 retail locations, shown in Figure 8.6, into six regions across the continental United States. They could assign the stores into standard geographic regions (e.g., Midwest, South, Northeast, etc.), but it's unlikely the company's stores will align with those predefined barriers. Instead, they attempt to let the data cluster itself into six regions with *k*-means. The dataset has 200 rows and two columns: latitude and longitude.[10]

---

[9] Lloyd, S. (1982). Least squares quantization in PCM. *IEEE transactions on information theory*, 28(2), 129–137.

[10] We are making many simplifying assumptions in this example. This approach would technically not be correct to group points on a sphere since latitude and longitude coordinates are not in Euclidean space. The distance metric we are using is ignoring the curvature of the Earth, as well as practical constraints, like access to highways.

FIGURE 8.6 The company's 200 locations, before clustering

The goal is to find six new locations on the map, each representing the "center" of the cluster. In numeric terms, this center point is essentially the average of every member in the group (that's the *means* of k-means). For this example, the centers could represent possible locations for regional offices, and each of the 200 stores would then be assigned to its closest office.

Here's how it works. To start, the algorithm selects six random locations to serve as potential regional offices. Why random? Well, it must start somewhere. Then, using the distance between points on our map (often described as, "as the crow flies"), each of the 200 stores is assigned to a cluster based on which of the six locations it's closest to. This is shown in Round 1 (top left) of Figure 8.7.

Each number represents the starting location and has an associated polygon that creates the border around the cluster. Notice in Round 1, that the "6" group is actually far away from its cluster—at least in this first iteration. Also notice that some of the randomly picked locations place the starting points into the ocean.

At each round of the algorithm, all points within a cluster are averaged together to create a center point (called a "centroid"), and the number moves to this new location. As a result, each of the 200 stores may now be closer to a different office than before. So, each store is reassigned to the regional location it's closest to. The process continues until points stop switching clusters. Figure 8.7 shows the k-means process after successive rounds of clustering.

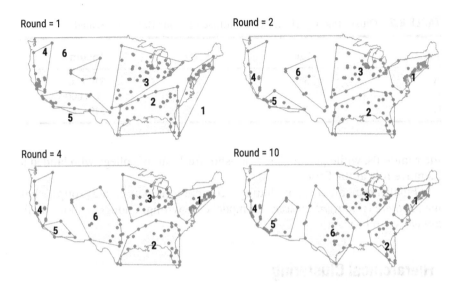

FIGURE 8.7 *k*-means in action on retail locations

With this insight, the company has clustered its 200 stores into six regions *and* identified possible locations within each cluster to serve as regional offices.

In summary, *k*-means tries to find any natural clusters that exist in the data and gradually pulls the *k* random starting points toward the clusters' centers like a magnet.

Potential Traps

We used "as the crow flies" distance in the previous example, but there are several types of distance formulas you could use when clustering a non-geographic dataset. It's beyond the scope of this book to identify all of them. Indeed, there is no right formula. However, you should not assume your analytics team used the best distance formula over the easiest. Make sure to ask which formula was used and why it was their best choice.

You also need to consider the scale of your data. You should not blindly trust the results because the math may group two things as "close" if they dominate the scale. For example, take three people at your workplace in Table 8.2. Which two would you say are "closest" to each other?

If the data isn't scaled appropriately, the income variable would dominate most distance formulas, an easy one being the difference (in absolute value) between any two data points. That means the "distance" between persons A and C would be "closer" than A and B based on income. That's even though persons A and B might be a better group reflecting two working parents in

TABLE 8.2 Clustering Algorithms Get Confused If Your Data Isn't Scaled.

Person	Age	Kids	Income
A	36	3	$100,000
B	37	2	$80,000
C	22	0	$101,000

their mid-30s, while person C is a hot-shot fresh out of college who landed a lucrative role at the firm.

Finally, remember we are letting a computer help us create groups. That means *there is no right answer*. All models are wrong. Though, if done well, *k*-means can be useful.

Hierarchical Clustering

Before finishing this section, we want to briefly touch upon hierarchical clustering, another popular clustering algorithm. In hierarchical clustering, the number of clusters isn't decided upon in advanced, like in *k*-means.

Think all the way back to the introduction of this chapter when you and a friend were organizing records without album covers. You didn't know how many clusters existed. You basically started with N groups: each record formed its own group. But as you listened, groups started to form naturally. Maybe you grouped two records in the category "contemporary jazz." If you had a group of three records in "classic jazz," you may decide it doesn't make sense to have such granular groups, so you combine the two groups into one "jazz" group.

Building up groups from the bottom-up like this will form a hierarchy in your data. Ultimately you decide at what level of the hierarchy you want to use to define your final groups.

CHAPTER SUMMARY

In this chapter, you learned about unsupervised learning, casually described as a way of letting data organize itself into groups. If you recall, we used this language to start the chapter but included a footnote saying it's not *quite* that easy. The ability to discover groups within data is a great power, but, as they

TABLE 8.3 Summarizing Unsupervised Learning and the Supervision Required

Unsupervised Learning	Dimensionality Reduction	Clustering
Example	Principal components analysis	k-means
What is it?	Grouping and condensing columns (features)	Grouping rows (observations)
What does it do?	Finds a smaller set of new, uncorrelated features that contain most of the information in the dataset	Groups similar observations together to create k meaningful "clusters" in your data
Why?	It lets you visualize and explore your data or reduces the dataset size to speed up computational time. PCA is usually an intermediary step in analysis.	It identifies patterns and structure in your data and gives you the ability to act on clusters differently (e.g., different marketing campaigns for different marketing segments).
Supervision Required	User must decide how to scale the data, how many principal components to keep, and how to interpret the principal components.	User must decide how to scale the data, what the right "distance" metric should be, and how many clusters to create.

say, with great power comes great responsibility. We hope you picked up on this underlying theme.

The ability to group data in any specific way is a product of the algorithm chosen, how it's implemented, the quality of the underlying data, and variation in the data. That means different choices can produce different groups. To put this bluntly, *unsupervised learning requires a lot of supervision*. It's not just a matter if hitting "go" on a computer and letting data organize itself. *You have decisions to make*, and we've summarized these for you (as well as the algorithms we discussed) in Table 8.3 for your reference.

To wrap up this chapter, we must reiterate that with unsupervised learning, there are no right groupings or right answers. You may, in fact, consider such exercises as a continuation of your exploratory data analysis journey described in Chapter 5. They help you see different views of the data.

CHAPTER 9

Understand the Regression Model

"Regression analysis is like one of those fancy power tools. It is relatively easy to use, but hard to use well—and potentially dangerous when used improperly."

Charles Wheelan in Naked Statistics[1]

SUPERVISED LEARNING

The preceding chapter showcased *unsupervised learning*—discovering patterns or clusters in a dataset without predefined groups. Remember, in unsupervised learning, we come with no preconceived notions or patterns. We instead take advantage of the underlying aspects of data, establish some guardrails, and then let the data organize itself.

However, there will be many occasions when you do in fact know something about the underlying data. In that case, you'll need to use *supervised learning* to find relationships in data with inputs and known outputs. Here, you have the correct answers to "learn" from. You can then judge the robustness of the model against what you know about the real world. A good model will allow you to make accurate predictions and provide some explanation to the underlying relationships between the data's inputs and outputs.

---

[1] Wheelan, C. (2013). *Naked statistics: Stripping the dread from the data.* WW Norton & Company.

Supervised learning, as you might recall, made an early cameo in the book's Introduction, back when you first started your Data Head journey. We asked you to predict if a new restaurant location would be a chain or independent restaurant. To make your guess, you first learned from existing restaurant locations (inputs) and known labels of "chain" or "independent" (outputs). You detected relationships between the inputs and outputs to create a "model" in your head that you used to make an educated prediction about the label for a new location.

You might be surprised to learn all supervised learning problems follow that same paradigm, depicted in Figure 9.1. Data with inputs and outputs, called *training data*, is fed into an *algorithm* that exploits the relationships between the inputs and outputs to create a *model* (think equation) that makes predictions. The model can take a new input and map it to a predicted output. When the output is a number, the supervised learning model is called a *regression* model. When the output is a label (aka categorical variable), the model is called a *classification* model.

You'll learn about regression models now and classification models in the next chapter.

This paradigm encapsulates many exciting and practical supervised learning problems found in technologies both old and new. The spam detector in your email, your home or apartment's estimated value, speech translation, facial detection applications, and self-driving cars—these all use supervised learning. Table 9.1 breaks down some of these applications into their inputs, outputs, and model types.

As the applications of supervised learning have become more advanced, it's easy to lose sight that these applications are rooted in a classic method

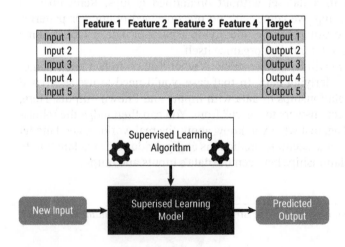

	Feature 1	Feature 2	Feature 3	Feature 4	Target
Input 1					Output 1
Input 2					Output 2
Input 3					Output 3
Input 4					Output 4
Input 5					Output 5

Supervised Learning Algorithm

New Input → Superised Learning Model → Predicted Output

FIGURE 9.1 Basic paradigm of supervised learning: mapping inputs to outputs

TABLE 9.1 Applications of Supervised Learning

Application	Input	Output	Model Type
Spam Detector	Email Text	Spam or Not Spam	Classification
Real Estate	Features and Location of House	Estimated Sales Price	Regression
Speech Translation	English Text	Chinese Text	Classification (each word is a label)
Facial Detection	Picture	Face Detected or No Face Detected	Classification
Smart Speakers	Audio	Did speaker hear "Alexa"?	Classification

created around 1800 called *linear regression*. Linear regression, specifically *least squares regression*,[2] is a workhorse in supervised learning and often the first approach your data workers will try when they want to make predictions. It's ubiquitous, powerful, and (don't be surprised) misused.

LINEAR REGRESSION: WHAT IT DOES

Suppose you run a small lemonade stand at the mall. You have a notion that temperature affects lemonade sales—specifically, the hotter it is outside, the more you'll sell. This prediction, if correct, could help you plan how much you'll need to buy and which days will sell more.

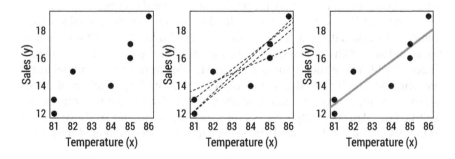

FIGURE 9.2 Many lines would fit this data reasonably well, but which line is the best? Linear regression will tell us.

---

[2] It's usually safe to assume "least squares regression" whenever you hear "linear regression," unless stated otherwise. There are other types of linear regression, but least squares is the most popular.

You plot some historical data, the left plot in Figure 9.2, and spot what appears to be a nice linear trend. If you fit a line through this data, you could use the equation[3] *Sales = m(Temperature) + b*. A simple equation like this is a type of model.[4] But how would you choose the numbers *m* (called the slope) and *b* (the intercept) to build your model?

Well, you could take a few stabs at it on your own. The center plot of Figure 9.2 shows four possible lines—these guesses all seem reasonable. But they're just guesses—they're not optimized to explain the underlying relationship given the data, even though they are close.

Linear regression employs a computational method to produce the line of *best fit*. By best fit, we mean that the line itself is optimized to explain as much linear trend and scatter of the data as possible. To the extent mathematically possible, it represents an optimal solution given the data provided. The rightmost chart in Figure 9.2 employs linear regression with the resultant equation: *Sales = 1.03(Temperature) – 71.07*.

Let's see how this works.

Least Squares Regression: Not Just a Clever Name

For a moment, let's just focus on our output variable—lemonade sales. If we wanted to figure out how much lemonade you might sell in the future, a reasonable prediction could be the mean of past sales (12 + 13 + 15 + 14 + 17 + 16 + 19) / 7 = $15.14. In short, we have an easy linear model, where *Sales* = $15.14.

Notice, this is still a linear equation, just without temperature. That means, no matter what the temperature is, we're going to predict sales of $15.14. We know it's naive, but it does fit our definition of a model that maps inputs to outputs, even if every output happens to be the same.

So how well does our simple model do? To measure the model's performance, let's calculate how far away the predicted sales value is from each of the actual sales. When the temperature was 86, sales were $19; the model predicted $15.14. When the temperature was 81, sales were $12; the model again predicted $15.14. The first instance represents an underprediction of about $4; the second, an overprediction of about $3. So far, we see that our model isn't perfect. And to really understand its ability to predict well, we'll need to understand the difference between what our model predicted and what

---

[3] In Algebra, you learned the equation of a line: $y = mx + b$. For any input x, you can get an output y by multiplying x and m and adding b. If $y = 2x + 5$, then an input $x = 7$ gives the output $y = 2 \times 7 + 5 = 19$.

[4] Quick terminology reminder: the output y is called the target, response variable, or dependent variable. The input x is called a feature, predictor, or independent variable. You may hear all terms in your work.

actually happened. These differences are called *errors*—and they represent how far off our prediction is from what we know to be true.

So how could we measure these errors to understand how well our model is doing? Well, we could take every actual sales price and subtract it from the mean, which we predicted to be $15.14. But if you do this, you'll see that every error, when summed up, always totals to zero. That's because the mean, which we're using as our predictor, represents the arithmetic center of these points. Subtracting the difference of all these points to the center makes the result zero.

But we do need a way to aggregate these errors because, clearly, the model isn't perfect. The most common approach is called the least squares method. And, in this approach, we instead *square* the differences from our prediction to make everything positive.[5] When we add up these numbers, the result won't be zero (unless, in fact, there is no error in our data). We call the result the sum of squares.

Let's take a look at Figure 9.3(a). You'll see the original scatter plot with *x* = *Temperature* and *y* = *Sales*. Notice, we've applied our naive model: temperature has no effect on the prediction. That is, the prediction is always 15.14, which gives a horizontal line at *Sales* = 15.14. In other words: the data point with temperature = 86 and actual sales = 19 has predicted sales of 15.14. The solid vertical line shows the difference between the actual value and predicted value for this point (and all other points). In regression, you square these lengths to create a square with area 14.9.

When sales were $15, our model still predicted $15.14, thus the corresponding square has an area of $(15.14 - 15)^2 = 0.02$. Add up the areas (sum the squares) from all points to represent the total error of your simple model. The squares you see on the right plot of Figure 9.3(a) are a visual representation of the squared error calculations. The larger the sum of squares, the worse the model fits the data. The smaller the sum of squares, the better the fit.

The question then becomes, can we arrive at a slope and intercept that would optimize (that is, reduce) sum of squared errors as much as possible? Right now, our naive model has no slope, but it does have an intercept of $15.14.

Clearly 9.3(a) doesn't do the best job. To get even closer to a good fit, let's add a slope *m*, which brings temperature into the equation. In Figure 9.3(b), we speculate a reasonable slope and intercept could be 0.6 and –34.91, respectively. Notice that adding this relationship makes our flat-lined model from 9.3(a) into a sloped line that appears to capture some upward trend. You can also immediately see a reduction in the overall area of the squares.

---

[5]Taking the absolute values would also make the errors positive before aggregating. However, squaring has nicer mathematical properties than the absolute value in that it is differentiable—this was vital to the early uses of linear regression because they had to do this all by hand.

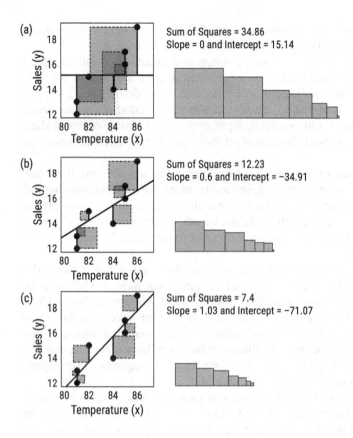

FIGURE 9.3 Least squares regression is finding the line through the data that results in the least squares (in terms of area) of the predicted values and the actual values.

Visually, the error has decreased substantially now that your model includes temperature. Your prediction for the point with temperature = 86 went from 15.14 in the simple model to $Sales = 0.6(86) - 34.91 = 16.69$, which reduced this observation's contribution to the sum of squares from 14.9 to $(16.69 - 19)^2 = 5.34$.

You could do the guesswork yourself—that is, plugging numbers into the slope and intercept until you might get the one combination that would reduce the sum of squares to its mathematical minimum. But linear regression does this for you mathematically. Figure 9.3(c) shows, quite literally, the least amount of squared errors you can see with this data. The slightest derivation from these slope and intercept numbers would make those squares bigger.

You can use the information to assess how well the final model fits the data. Note Figure 9.3(c), which contains the linear regression result, is still imperfect. But you can clearly see it's better than Figure 9.3(a) where you predicted $15.14 each time.

How much better? Well, you started with an area (sum of squares) of 34.86, and the final model reduced the area to 7.4. That means we've reduced the area of the control model by (34.86 – 7.4) = 27.46, which is a 27.46/34.86 = 78.8% percent reduction in total area. You'll often hear people say the model has "explained," "described," or "predicted" 78.8% (or 0.788) of the variation in the data. This number is called "R-squared" or R^2.

If the model fits the data perfectly, $R^2 = 1$. But don't count on seeing such models with a high R^2 in your work.[6] If you do, someone probably made a mistake—and you should ask to review the data collection processes. Recall from Chapter 3 that variation is in everything, and there's always variation we can't fully explain. That's just how the universe is.

LINEAR REGRESSION: WHAT IT GIVES YOU

Let's quickly review what we've discussed and tie this back to the supervised learning paradigm in Figure 9.1. You had a dataset with one input column and one output column that was fed into the linear regression algorithm. The algorithm learned from the data the best numbers to finalize the linear equation *Sales = m(Temperature) + b* and produced the model *Sales* = 1.03(*Temperature*) – 71.07 that you can use to make predictions about how much money you'll make selling lemonade.

Linear regression models are favorites in many industries because they not only make predictions, but also offer explanations about how the input features relate to the output. (Also, they're not hard to compute.) The slope coefficient, 1.03, tells you for every one unit increase in temperature, you can expect sales to increase by $1.03. This number provides both a magnitude and direction for the impact of the input on the output.

Recognizing there is randomness and variability in the world and the data we capture from it, you can imagine that the coefficient numbers from linear regression have some variability built in. If you collected a new set of data from your lemonade stand, perhaps the impact from temperature would change from $1.03 to $1.25. The data fed into the algorithm is a sample, and thus you must continue to think statistically about the results. Statistical

---

[6] For simple regression with one input, R^2 is the square of the correlation coefficient we discussed in Chapter 5. It is possible, however, for R^2 to be negative. This happens when the linear regression model is worse than predicting the mean.

software helps you do this by outputting p-values for each coefficient (testing the null hypothesis, H_0: the coefficient $= 0$), letting you know if the coefficient is statistically different than zero. For instance, a coefficient of 0.000003 is very close to zero, and might reflect something that for practical purposes is a zero in your model.

Put differently, if the coefficient is not statistically different than zero, you can drop that feature from your model. The input is not impacting the output, so why include it? Of course, statistics lessons from Chapter 6 remain relevant. A coefficient may indeed be statistically significant but not practically so. Always ask to see the coefficients of models impacting your business.

Extending to Many Features

Your business, we assume, is a little more complicated than a lemonade stand. Your sales would not only be a function of temperature (if a seasonal business), but many other features or inputs. Fortunately, the simple linear regression model you've learned about can be extended to include many features.[7] Regression with one input is called simple linear regression; with multiple inputs, it's called multiple linear regression.

To show you an example, we fit a quick multiple linear regression line to the housing data we explored in Chapter 5. This data had 1234 houses and 81 inputs, but to simplify the example, we're going to look at 6 inputs. (We could have also used PCA to reduce the dimensionality, but we didn't want to overcomplicate this example.)

TABLE 9.2 Multiple Linear Regression Model Fit to Housing Data. All corresponding p-values are statistically significant at the 0.05 significance level.

Input	Coefficient	p-value
(Intercept)	−1614841.60	<0.000
LotArea	0.54	<0.000
YearBuilt	818.38	<0.000
1stFlrSF	87.43	<0.000
2ndFlrSF	90.00	<0.000
TotalBsmtSF	53.24	<0.000
FullBath	−7398.13	0.017

---

[7]The upper limit for the number of features/inputs in a linear regression model is $N - 1$, where N is the number of rows in your dataset. Thus, you can have up to 11 inputs to predict monthly sales for a 12-month period.

Let's build a model to predict the sales price of a home (output) based on lot area, year built, 1st floor square footage, 2nd floor square footage, basement square footage, and number of full bathrooms. The linear regression algorithm will work its magic, learn from the data, and output the best coefficients to complete the model, resulting in the intercept and coefficients in Table 9.2.

A core tenet of a multiple regression model is to isolate the effect of one variable while controlling for the others. We can say, for example, if all other inputs are held constant, a home built one year sooner adds (on average) $818.38 to the sales price. The coefficients of each feature are showing the magnitude and direction the input has on price. Make sure to consider the units. Adding 1 unit to square footage is different than adding 1 unit to the number of bathrooms. A statistician can scale these appropriately if you need an apples-to-apples comparison between coefficients.

Each coefficient also undergoes an associated statistical test to let us know if it's statistically different than zero. If it is not, we can safely remove it from the model because it doesn't add information or change the output.

LINEAR REGRESSION: WHAT CONFUSION IT CAUSES

If your esteemed authors were con men, we would have ended the chapter with the previous section and asked you to purchase linear regression software as the cure-all for your business needs. Our sales pitch would be "Enter data, get a model, and start making predictions about your business today!" That sounds fantastically easy, but by this point you've gotten to know data well enough that nothing is as easy at seems (or as easy as it's sold). Like the quote we used to start the chapter, linear regression in the wrong hands can be potentially dangerous. So whether you create or use regression models, always maintain a healthy skepticism. The equations, terminology, and computation make it seem like a linear regression model can autocorrect any issue in your dataset. It can't.

Let's go through some of the pitfalls of using linear regression.

Omitted Variables

Supervised learning models cannot learn the relationship between an input variable and an output variable if the input variable was omitted from the model. Consider our simple model that predicted lemonade sales based on average past sales but without regard to temperature.

Data Heads like yourself, now aware of this issue, might suggest informative, relevant features to include in models. But don't just leave feature selection up to your data workers. Subject matter expertise and getting the right data into a supervised learning model is key to having a successful model.

For example, the housing model in the previous section has an R^2 value of 0.75. This means we've explained 75% of the variation in sales price with our model. Now think about features *not* in this model that would help one predict a home's price—perhaps things like economic conditions, interest rates, elementary school ratings, etc. These omitted variables not only impact predictions of the model but may lead to dubious interpretations. Did you notice in Table 9.2 the coefficient for the number of bathrooms was negative? That doesn't make sense.

Here's another example. Consider a linear regression model that "learned" shoe size has a large positive coefficient in a linear regression model for the number of words a person can read per minute. Age is clearly a missing input variable here: including it would remove the need for "shoe size" as an input. Certainly, examples in your work may not be as easy to tease out as this example but believe us when we say omitted variables can and will cause headaches and misinterpretations. Further, many things are correlated with time, a common omitted variable.

We hope the phrase "correlation does not mean causation" comes to mind once again while reading this section. Having one variable labeled as an input to a model and another as an output does not mean the input *causes* the output.

Multicollinearity

If your goal with linear regression is interpretability—to be able to study the impact of the input variables on the output by studying the coefficients—then you must be aware of multicollinearity. This means several variables are correlated, and it creates a direct challenge to your model being interpretable.

Recall, a goal of multiple regression is to isolate the effect of one input's contribution while holding every other input constant. But this is only possible *if the underlying data is uncorrelated.*

To offer a quick example, suppose the lemonade stand data we plotted earlier had the temperature in both Celsius and Fahrenheit. Obviously, the two are perfectly correlated because one is a function of the other, but for this example, suppose each temperature reading was taken with a different instrument to introduce some variation.[8] The model would go from:

1. *Sales* = 1.03(*Temperature*) – 71.07 to
2. *Sales* = –0.2(*Temperature in F*) + 2.1(*Temperature in C*) – 30.8.

---

[8] Linear regression models will not compute if two inputs are perfectly correlated, so we added noise to this example.

Now it looks like adding a degree in Fahrenheit has a negative impact! Yet, we know the inputs are comingled—in fact, redundant—and linear regression cannot break apart the relationship. Multicollinearity is present in most observational datasets, so consider this a fair warning. Data collected experimentally, however, is specifically designed to prevent the inputs from being multicollinear to the extent possible.[9]

Data Leakage

Imagine building a model to predict the sales price of a home (like we did earlier in the chapter). But this time, the training data not only includes features about the home (size, number of bedrooms, etc.), but also the initial offer made on the house. A snapshot of the dataset is shown in Table 9.3.

If you run the model on the data, you might notice that the initial offer is very predictive of the sales price. *Great!*—you think. You can rely on that to help you predict home prices for your company.

The model then goes into production. You attempt to run the model only to find that you don't have access to the initial offer on the houses whose sales prices you are trying to predict—they haven't sold yet! This is an example of data leakage. Data leakage[10] happens when a concurrent output variable masquerades as an input variable.

The problem with using the initial offer is one of timing. Consider for a moment that you could only learn what the starting offer was *after* the home was already sold.

As we are immersed in data, data leakage is easy to overlook. Sadly, many textbooks don't cover it because the datasets inside are pristine for learning, but real-world datasets always have the possibility for leakage. As a Data Head, you must be on the lookout to ensure your input and output data do not contain overlapping information.

We will explore data leakage again in later chapters.

TABLE 9.3 Sample Housing Data

Square Footage	Bedrooms	Baths	Initial Offer	Sales Price
1500	2	1	$190,000	$200,000
2000	3	2	$240,000	$250,000
2500	4	3	$300,000	$300,000

---

[9] There is an entire field of Statistics called Design of Experiments devoted to this idea.

[10] en.wikipedia.org/wiki/Leakage_(machine_learning)

Extrapolation Failures

Extrapolation means predicting beyond the range of the input data you used to build your model. If the temperature was 0°F, the lemonade stand model would predict sales of –$71.07. If a house had no square footage and no bathrooms (practically speaking, the house wouldn't exist), the model would predict a sales price of –$1,614,841.60. Both cases are nonsense.

The models are predicting outside the range of data they "learned" from, and unlike humans, equations don't have common sense to know it's wrong. Math equations cannot think. If you give it numbers as inputs, it will spit back a number as an output. It's up to you, the Data Head, to know if extrapolation is taking place.

It must be stressed that model results are always predicted on the given data. Which is to say, you should not make predictions with data that "fits" the range of the training data but not the context in which the dataset was collected. The model has no context for changes taking place in the world.

If you built a model that predicted housing prices in 2007, your model would have performed terribly in 2008 after the housing market crashed. Using the model in 2008 would have extrapolated the market conditions of 2007 data to be vastly different in 2008. Industries are facing this at the time of writing, 2021, because of the COVID-19 pandemic. Models trained on pre-COVID data no longer reflect many of the relationships identified, and the models are no longer valid.

Many Relationships Aren't Linear

Linear regression would not do well modeling the performance of the stock market. Historically, it has grown exponentially, not linearly. The Statistics department at Procter & Gamble would offer this advice: "Don't fit a straight line through a banana."

Statisticians have tools in their arsenal to transform some non-linear data into linear data. Even so, sometimes, you must accept that linear regression is not the right tool for the job.

Are You Explaining or Predicting?

Throughout this chapter, we've been discussing two possible goals with regression models: explain relationships and make predictions. Linear regression models can seemingly do both. A linear regression model's coefficients (under the right conditions) offer interpretability. Many industries focus their attention on interpretability—like in clinical trials where researchers need to know the precise magnitude and direction of the input "dosage of a medicine" to the output "blood pressure." In this case, great care must be taken to avoid multicollinearity and omitted variables to ensure model explanations are possible.

In other fields, like machine learning, accurate prediction is the goal.[11] The presence of multicollinearity, for example, may not be a concern if the model can predict future outputs well. When the purpose of a model is to predict new outputs, you absolutely must be careful to avoid *overfitting* the data.

Recall that models are simplified versions of reality. A good model approximates the relationships between inputs and outputs well. Indeed, it's capturing some underlying phenomenon. The data itself is merely an expression of this underlying phenomenon.

An overfit model, however, does not capture that relationship we speculate exists. Rather it captures the interaction of the training data, including the noise and variation in the data itself. Thus, the predictions it makes are not what's being modeled, but rather just the data points we have.

Models that overfit effectively memorize the sample training data. They don't generalize well to new observations. Check out Figure 9.4. On the left, you see the lemonade data with the linear regression model. On the right, a complex regression model that perfectly predicts some of the points. Which would you want to use to make a prediction?

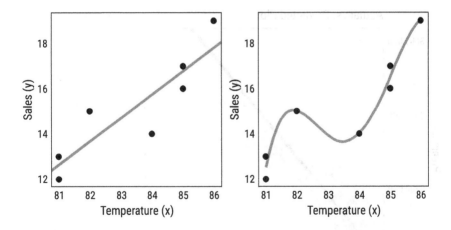

FIGURE 9.4 Two competing models. The model on the left generalizes well, while the model on the right overfits the data, effectively memorizing the data. Because of variation, the model on the right will not predict new points well.

---

[11] There's a great paper that goes into depth about the difference between explaining and prediction for models: Shmueli, G. (2010). To explain or to predict? *Statistical science, 25*(3), 289–310.

The way to prevent overfitting is to split a dataset into two parts: a *training set* you use to build a model and a *test set* to see how it performs. The performance on your test set, data your model did not learn from, will be the ultimate judge of how well your model predicts.

Regression Performance

Whenever you come across a regression model at work, whether it's a multiple linear regression model or something more sophisticated that was just invented last week, the best way to judge how well it fits your data is with an "actual by predicted" plot. In our experience, some people assume we can't visualize regression performance when we have too many inputs, but remember what the model has done. It converted inputs (either one or several) into an output.

So, for each row in your dataset, you have the actual value and its associated predicted value. Plot them in a scatter plot! These should be well correlated. This eye-ball test will let you know quickly how well your model has done. Your data workers can provide several associated metrics (R-squared being one), but never look at those numbers alone. Always, *always* demand to

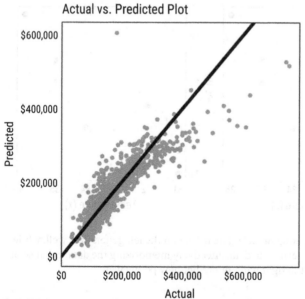

FIGURE 9.5 In this plot, you can see how the model does not do well predicting high-value homes. Can you spot other issues with this model using this chart?

see the actual vs. predicted plot. An example is shown in Figure 9.5 using the model we built on the housing data.

OTHER REGRESSION MODELS

Finally, you may come across some variants of linear regression called LASSO and Ridge Regression that help when there are many correlated inputs (multicollinearity) or when you have more input variables than rows in your dataset. The result is a model that looks similar to multiple regression models.

Other regression models look completely different. K-nearest neighbor from the Introduction was applied to a classification problem, but it can easily be applied to a regression problem. For instance, to predict the sales price of any home, we could take the mean selling price of the three closest homes to it that recently sold. This would use K-nearest neighbor as a regression problem.

We'll go over some of these models in the next chapter, as they can be used for both regression or classification.

CHAPTER SUMMARY

Our goal was to help you develop an intuitive understanding of supervised learning and its most fundamental algorithm, linear regression. We then examined the many ways regression models can go wrong. Keep those issues and pitfalls in your mind because it's not a matter of *which* will impact regression models in your workplace, but *how many*.

As you might have put together by now, supervised learning is both powered and limited by the training data. Unfortunately, we see companies spending more time worried about the latest supervised learning algorithm rather than thinking about how to collect relevant, accurate, and enough data to feed the algorithm. Please, do not forget about the "garbage in, garbage out" mantra from earlier in the book. Good data is key and is the lifeforce for supervised learning models.

And at this point, we can reveal that if you understood unsupervised learning and supervised learning, you now understand machine learning. Congrats! And apologies for the lack of a big reveal. We wanted to be the antithesis of a sales pitch and teach you components of machine learning without getting bogged down by hype. Machine learning *is* unsupervised learning and supervised learning.

Let's continue this conversation about machine learning in the next chapter with classification models.

CHAPTER 10

Understand the
Classification Model

A Machine Learning algorithm walks into a bar. The bartender asks, "What'll you have?" The algorithm says, "What's everyone else having?"

Chet Hasse (@chethaase)

In the previous chapter, we described supervised learning with *regression models*. Specifically, regression models allow us to predict values—that is, a number, like sales—by fitting a model to our feature set. But what if you were attempting to predict a specific outcome—like will a person with a certain set of demographic features want to buy a book about data? Indeed, if you've ever wondered how companies predict if you'll click a specific ad, buy a product (and which product you'll buy), default on your car loan, get a job interview, or develop a disease, this chapter will be your guide.

Such problems—where there is a categorical variable (i.e., a label) to be predicted—employ *classification models*.

INTRODUCTION TO CLASSIFICATION

Classification models can predict two outcomes, called *binary classification*; or multiple classes, called *multiclass classification*.[1] Predicting if someone

---

[1] Be careful not to confuse clustering with classification. Recall, there are no labels in clustering. In clustering, if labels are assigned, it happens by you, the analyst, *afterward*. In classification, we begin with labels in a dataset.

defaults on a car loan is binary (yes/no), while predicting which car someone will buy is multiclass (Honda, Toyota, Ford, etc.). For simplicity, we'll focus on binary classification problems. However, additional classes are a natural extension of the topics we'll discuss.

The outcomes of some classification models are often described as being "positive" or "negative." Remember, science likes to test for things in the affirmative. As a result, you should construe positive and negative as meaning "do" and "does not," respectively. It's a way of separating observations that *demonstrate an action* (click, buy, default, have a disease) from those that don't. In others, say predicting a voter's political party affiliation, you want to be clear which class is "positive" or "negative" in modeling vernacular to avoid any confusion. For instance, how you assign the voter affiliation of Democrat or Republican as positive or negative in your model is not a commentary on either, but rather an arbitrary labeling. As a Data Head, you should ensure everyone on the team is aligned with the model's assignment.

What You'll Learn

In this chapter, we'll use a human resources dataset to describe the following classification models:

- Logistic regression
- Decision trees
- Ensemble methods

Logistic regression[2] and decision trees are among the most taught in data science curricula and are heavily accommodated by software programs. Their ease of use and interpretability make them an ideal choice for some problems. However, like all algorithms described in this book, they're not without their own set of pitfalls.

We'll also introduce you to ensemble methods, which are becoming a new standard for data workers, especially in data science competitions.[3]

In the second half of this chapter, we'll go into more detail about *data leakage* and *overfitting*. We'll leave a discussion of accuracy to its own section at the end of the chapter, as the term (in the context of data) requires a hefty dose of nuance to really understand. And we don't want you to make the same mistakes we've seen others make time and again.

---

[2] Logistic regression, as you will learn, predicts probabilities. By adding a decision rule, it becomes a classification algorithm.

[3] As a point of clarification, decision trees and the ensemble methods we introduce can be applied to regression problems. So, if your dataset's output is a number, give these a try.

Classification Problem Setup

Suppose every summer, hundreds of college students apply for a data science internship at your company. Reviewing hundreds of applications by hand would be a tough task to do. Could you find a way to automate this process?

Fortunately, your company has a stockpile of historical data to learn from—information about each applicant and a yes/no label to indicate if they were invited for an interview. With historical data and a classification model like logistic regression, you could develop a predictive model that takes data from a person's online application as inputs—attributes such has grade point average (GPA), year in school, major, number of extracurricular activities—and outputs a prediction if the applicant should be offered an interview. If it works, there's no need to review resumes by hand.

How could you solve this problem? We'll start with logistic regression.

LOGISTIC REGRESSION

Let's start simple by reviewing data from ten past applications using only their GPA as an input. Because computers only understand numbers, you can convert "yes" and "no" to 1 and 0, respectively, with 1 being the positive class. The data is shown in Table 10.1. Looking down the table, the general trend is not surprising: students with a higher GPA are more likely to be offered an interview.

If you tried to implement linear regression on this data (like we learned about in the previous chapter), you would get strange results. For instance, if

TABLE 10.1 Simple Dataset for Logistic Regression: Using GPA to Predict Interview Offer

Application Id	GPA	Offer (yes/no)	Offer (1/0)
1	2.00	no	0
2	2.20	no	0
3	2.50	no	0
4	2.80	yes	1
5	2.85	no	0
6	3.50	yes	1
7	3.60	no	0
8	3.70	yes	1
9	3.80	yes	1
10	4.00	yes	1

we plug Table 10.1 into statistical software and generate the regression model, we would get an equation that looks like this:

$$Offer = (0.5) * GPA - 1.1.$$

But let's think about this model for a second. Suppose a new applicant has a 2.0 GPA. The regression model would output *Offer* = (0.5)*(2.0) – 1.1 = –0.1. An applicant with a 4.0 GPA would have an output 0.9. But what do the numbers –0.1 and 0.9 mean in the context of predicting whether this candidate was extended an offer? (Hint: we're not sure either.)

What would be helpful would be a prediction about the *probability* of an applicant being offered an interview based on their GPA. For example, you might learn applicants with 2.0 GPAs had a 4% probability of being offered an interview while applicants with 4.0 GPAs had a 92% probability. This information is relevant to the task at hand because it could help you establish business rules on how future applicants should be classified. But remember probabilities must fall within 0 and 1, inclusive, and regression models do not operate within these constraints. Regression models are unbounded, able to output any value you can imagine. So clearly linear regression isn't the right choice for this problem.

Thus, you need a way to constrain the output from an equation of the form $y = mx + b$ to ensure it lies in the range of proper probabilities. This is precisely what logistic regression does: it "squishes" numbers to guarantee model outputs always lie between 0 and 1, giving the user a predicted probability of belonging to the positive class (in this case, offer = "yes").

Look at the equation for logistic regression:

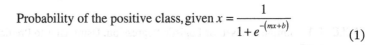

$$\text{Probability of the positive class, given } x = \frac{1}{1 + e^{-(mx+b)}} \tag{1}$$

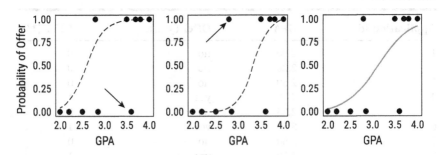

FIGURE 10.1 Fitting different logistic regression models to the data. The model on the right is the best fit.

Does the *mx+b* part look familiar? Yes, it's formula from linear regression. It's now embedded within this equation, called the logistic function (hence the name logistic regression).[4] This function makes sure the resulting number is a probability.

Let's look at some plots to make this clearer. Figure 10.1 has three scatter plots of the data in Table 10.1. (We plotted the "line of best fit" in the previous chapter with a very similar set of three charts.) Each of these plots reflects a different set of input values for *m* and *b* in equation 1. Recall that in linear regression, the values for *m* and *b* modulated the perfect position for the line that reduced error as measured by the sum of squares, but we've established that a straight line from linear regression won't fit this data well; the line would extend below 0 on the left and 1 on the right. The equation in (1), however, regardless of the values of *m* and *b*, will always produce an S-like curve that lies between 0 and 1.

Start with the left and middle plots in Figure 10.1 and spot their weaknesses. The left plot shows a model (the dashed line) that is overly confident predicting that a high GPA results in an interview; it has a huge miss with the 3.5 GPA applicant who was denied. In the middle plot, the model is giving an unreasonably low probability to students with a low GPA. The student with a 2.8 GPA, who was invited for an interview, was given a near zero chance by this model. The rightmost plot in Figure 10.1 shows the best balance. It's the output from the logistic regression algorithm, which strikes a nice equilibrium between the left and middle charts. As it turns out, it's the mathematically most optimized solution that exists for the data points present. That resulting logistic regression model has the following equation:

$$\text{Probability of getting an offer, given a GPA} = \frac{1}{1+e^{-\left(2.9*GPA-9.0\right)}} \tag{2}$$

Logistic regression reduces what's called *logistic loss*. Logistic loss is simply a way of measuring how close the predicted probabilities are to the actual labels. Though linear and logistic regression use different methods, they are of the same statistical spirit—make all predicted values from a model as close as possible to the actual values (in the aggregate).

---

[4]The number *e* in the equation is a mathematical constant, like π, that has many important applications, logistic regression being just one. Known as Euler's constant, *e* approximately equals 2.71828.

Logistic Regression: So What?

Logistic regression provides two benefits: we get a formula that helps make predictions informed by data; and, the coefficients of that formula explain the relationships between the inputs and outputs.

Here's how you could apply it. Figure 10.2 shows the probability of an offer for a 2.0 GPA student from our logistic regression model. That person would have about a 4% chance at an offer. An applicant increasing their GPA from a 2.0 to a 3.0 improves the probability of an interview from 4% to 41%—a 37% difference. But increasing from a 3.0 to a 4.0, still a one-unit increase, swings the probability from 41% to 92%, a 51% difference! Note: the impact on probability of an extra GPA point is not constant in logistic regression models. This is another way in which logistic regression differs from linear regression: in linear regression, increasing one unit of the input variable has the same effect on the output, no matter the starting value.

You'll notice that logistic regression, by itself, can't tell you whether someone should be made an offer or not. Rather, it provides you with a probability of an offer. If you want to automate a decision with logistic regression, you'll need to set a *cutoff*, also called a decision rule. The cutoff probability will affect how you implement what you've learned. If you set the cutoff to 90%—that is, you'll only want to look at applications whose GPA's suggested a 90% probability of receiving an offer previously—you'll likely extend fewer offers. On the other hand, if you are happy with seeing applications where past data provides them a 60% chance of an offer, you'll see far more candidates. Setting things like cutoffs requires the contributions of domain experts.

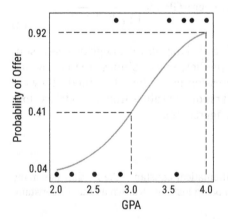

FIGURE 10.2 Applying the logistic regression model to make predictions at GPA = 2, GPA = 3, and GPA = 4

As we also said previously, the coefficient of any regression functions tells us about the relationships between the inputs and output. At a glance, we can see if the coefficient for GPA is positive in equation (2); its value is 2.9. That lets us know that a higher GPA improves one's chances. Not exactly earth-shattering news in this case, but for researchers predicting if someone might develop cancer based on certain biomarkers, this could matter a great deal.[5]

What to Watch Out for When Working with Logistic Regression

Logistic regression models inherit the concerns and confusions we detailed in the previous chapter for linear models. Specifically:

- **Omitted Variables:** An algorithm cannot learn from data that isn't there.
- **Multicollinearity:** Correlated input features can widely mess up your interpretation of model coefficients, sometimes changing a coefficient from positive to negative (or vice versa).
- **Extrapolation:** Extrapolation is slightly less of a concern in logistic regression than linear regression because the outputs will never fall outside the range 0 to 1. But don't get too comfortable. Predicting beyond the range of the training data may result in overly confident probabilities because those predictions become asymptotically closer to one.

To be sure, there are other mistakes to avoid in logistic regression, and we'll return to them at the end of the chapter.

DECISION TREES

Some people are turned off (afraid, perhaps) by the math required by logistic regression. Plus, not every relationship between inputs and outputs follows the linear mold of $y = mx + b$. An alternative approach, one that is easily

---

[5]To really unpack the formula, you'll need an understanding of *log odds*. A greater understanding of log odds, however, is beyond the scope of this book.

TABLE 10.2   Snapshot of the Intern Dataset from HR. The majors are CS = Computer Science, Econ = Economics, Stats = Statistics, and Bus = Business.

Student ID	GPA	Year	Major	Number of Extracurriculars	Offer?
1	3.41	1	CS	1	No
2	3.33	3	Econ	2	No
3	2.96	3	CS	5	Yes
4	3.28	2	Stats	4	Yes
5	2.78	2	CS	3	No
6	3.01	4	Econ	0	No
7	2.56	3	Stats	2	No
8	2.72	3	CS	4	Yes
9	2.00	3	Stats	2	No
10	2.42	1	Bus	3	No
⋮	⋮	⋮	⋮	⋮	⋮

digestible and simple to visualize, is a decision tree. Decision trees will split a dataset into multiple parts, providing a list of rules to guide your predictions like a flowchart.

Take, for instance, the dataset in Table 10.2. Here you see a sample of 10 students (out of 300) who have applied and been offered interviews at your company. Rather than using GPA as the sole input into your process, you decide to analyze all features to explore how interview offers were made in the past. Note, in this dataset, 120 students (40%) were offered an interview.

If you wanted to use these features to help understand who received an offer and who didn't, you might theorize a few rules on your own: students with a high GPA who are involved in extracurriculars probably have a higher chance of getting an offer. But what GPA value would you use to "split" your students? 3.0? 3.5? And what information would you share to justify your decision? You see what we're implying: generating rules on your own would be a monumental task. Fortunately, a decision tree algorithm can do this for you. It searches for the input feature—and the value of that feature—that best separates the students who received an offer from those who didn't. And then it finds the next feature that helps separate those two groups at a more granular level, and on and on.

We ran our dataset through a decision tree algorithm known as CART[6] and generated the decision tree in Figure 10.3. It's more of an upside-down

---

[6] There are several algorithms to generate decision trees, the most popular being CART (Classification and Regression Trees). For more on CART, see Breiman, Leo; Friedman, J. H.; Olshen, R. A.; Stone, C. J. (1984). *Classification and regression trees*. Monterey, CA: Wadsworth & Brooks/Cole Advanced Books & Software.

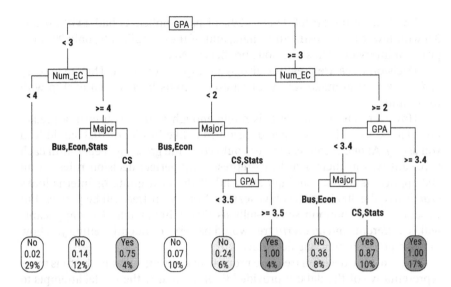

FIGURE 10.3 Simple decision tree applied to the HR intern dataset

tree, made of decision "nodes," "branches," and "leaves," where the final pre-diction is determined by a leaf. Let's traverse the tree with an example observation to show how it works.

Suppose an applicant, call her Ellen, is a sophomore with a 3.6 GPA majoring in Computer Science and participates in a club sport. In data terms, she is encoded as {GPA = 3.6, Year = 2, Major = CS, Num_EC = 1}, where Num_EC means "Number of Extracurriculars."

At the top of Figure 10.3 is the root node, which tells us the feature that best splits the data: GPA. Ellen has a 3.6 GPA, so she moves to the right branch into the next decision node: Num_EC. Her Num_EC is 1, so she moves to the left branch into the next decision node: Major. CS majors move to the right, and then again GPA enters the process. Her GPA is at least a 3.5, so you'd predict "Yes," she'd be offered an interview.

Notice how the tree uncovers interactions in the input features. Here, not being involved in many extracurriculars is offset by having a high GPA in CS or Statistics.

The numbers in the leaves, at the bottom of Figure 10.3, summarize how the decision tree split the training data. The leaf on the far right has three data points: {Yes, 1.00, and 17%}. This tells us that 100% of past applicants with a GPA ≥ 3.4 and at least 3 extracurriculars were offered an interview, regardless of major. (You can see this if you trace the leaf back up to the root!) Any new applicant fitting this description would have the prediction = "Yes" because the percentage of past applicants in this leaf were above 50%. This represents 17% of the training data (51 applicants).

The leaf on the far left tells us 29% of past applicants had a GPA below 3.0 with fewer than 4 extracurriculars, and of these applicants, only 2% were given an interview. Thus, the node predicts a "No."[7]

Decision trees are great for displaying exploratory data. They're an easy and quick way to make sure your dataset's inputs have a relationship with the output.

However, one tree by itself is rarely enough to help you form a prediction. Think about how a single decision tree, like in Figure 10.3, could lead you astray. At one extreme, the tree could continue growing deeper until each applicant was in their own leaf—representing perfect decision rules for all 300 applicants in the training data. And, if the next group of interns looks *exactly* like the 300 interns you've learned from, your tree will be perfect. But variation—and common sense—tells us this is impossible. New applicants will be different, and an *overfit* tree would be very confidently telling you how to make potentially wrong decisions.

Indeed, lone decision trees are prone to overfitting—that is, the fit is more representative of the dataset provided than the reality the model attempts to predict. One way to fix this is through "pruning," but lone trees remain very sensitive to their training data. If you were to sample 100 applicants from your data and build a new decision tree, you'd likely find different decision nodes and split values within the tree. The root node, for example, may split at a 3.2 GPA instead of 3.0.

How could you fix these issues with decision trees? Enter ensemble methods.

ENSEMBLE METHODS

Ensemble methods, so named because they represent the aggregation of different results by running an algorithm dozens, perhaps even thousands, of times are popular among data scientists because of their ability to make meaningful predictions at a granular level.

Two methods in particular, *random forests* and *gradient boosted trees*, have emerged as data scientists' favorites. They're often used by the winning teams in data science competitions on the website Kaggle.com, where companies post datasets and challenge data scientists to build the most accurate models—offering hefty prize money to the champions. In this section, we'll give you a brief, intuitive explanation of these esteemed methods.

---

[7] We generated this tree and visual using the open source (free) statistical software R, along with the "rpart" and "rpart.plot" packages. Not all decision trees you see will display this level of detail.

Random Forests

Take any two experienced human interviewers and here's what you'd notice—each has their own internal decision rules based on their individual past experiences and the types of applicants they've interacted with. Said simply, they'll evaluate candidates differently. Which is why, in many companies, hiring is a team effort—the decision is the consensus from several employees' individual assessments to balance out any glaring differences in any one individual's decision making.

A random forest[8] is the decision tree equivalent to this idea. The algorithm takes a random sample of data and builds a decision tree. And then it repeats that process, several hundred times more.[9] The result is a "forest" of trees that mimic many independent evaluators of your dataset, with the final prediction being the consensus from the trees by majority vote. (Random forests can also output the average of predicted probabilities for classification problems or the average of continuous numbers in regression problems.)

Figure 10.4 shows four trees in our forest. Look closely, and you'll notice another feature of random forests. In two of the trees, GPA is the first split. In another, Major, and finally, the number of extracurriculars. This is by design. Random forests not only randomly select which observations (rows) to build a tree with, but they also randomly selects the features (columns). This decorrelates the trees in the forest, allowing each to find new interactions in the data. Otherwise, the trees would find redundant information.

Gradient Boosted Trees

Gradient boosted trees[10] take a different approach. Whereas a random forest creates hundreds of individual trees and averages their outputs at the end, gradient boosted trees build trees *sequentially*.

The human resource equivalent of gradient boosting is having multiple interviewers line up sequentially out the door to interview a candidate in succession. Each interviewer would enter the room, ask the candidate one or two questions, leave the room, and tell the next interviewer something like, "So far, I would hire this person, but we need to ask more probing questions

---

[8] Breiman, L. (2001). Random forests. *Machine learning, 45*(1), 5–32.

[9] Building models on random samples of your data is known as "bagging." Random forests are a particular application of bagging.

[10] See: Friedman, J., Hastie, T., & Tibshirani, R. (2001). *The elements of statistical learning* (Vol. 1, No. 10). New York: Springer series in statistics, Chapter 10 and the references therein for more on boosting. Be advised, this is an advanced text.

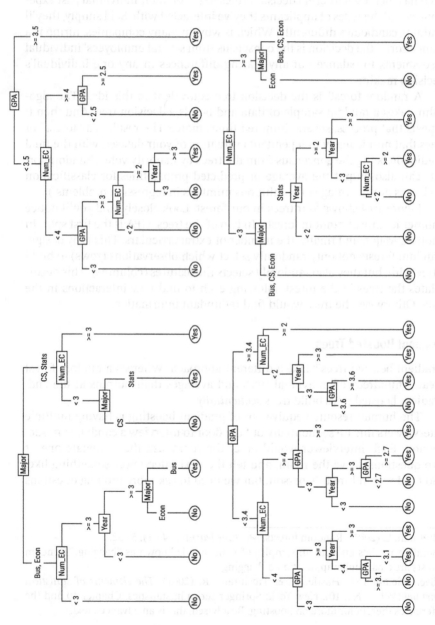

FIGURE 10.4 A random forest is a "forest" of several decision trees, usually hundreds, where each decision tree is built on a random subset of the data. The final prediction for an observation is the consensus of the trees in the forest.

in these areas," and so on down the line. The result is one recommendation, built and tweaked in a cumulative fashion from the group, rather than many separate recommendations that are aggregated into one.

Gradient boosted trees typically start by building what's called a "shallow" tree—a tree with few branches and nodes. This is inherently naive—it's the first iteration. And it doesn't do a good job of splitting the dataset into a correct classification. In the next step, a new tree is built on the *errors* of the first tree, giving it a *boost* on observations where the errors were large (that's the *gradient* at work). The process continues, potentially thousands of times on larger datasets, to create a boosted model.

Generally, these ensemble methods are not for "small data," so data workers should apply them when they have hundreds, not dozens, of observations.

Interpretability of Ensemble Models

Consider trying to make sense of thousands of tree leaves and nodes whose rules are sensitive to minute changes in the data. These models are often called "black boxes" because it's hard to understand their inner workings. What you gain in accuracy with random forests and gradient boosted trees over logistic regression, you lose in interpretability. It's a tradeoff.[11]

We'll discuss other black box models in Chapter 12, "Conceptualize Deep Learning."

WATCH OUT FOR PITFALLS

As powerful as classification is, the potential for its misapplication is significant—there are several traps to fall into. Make no mistake, models that suffer from the pitfalls described in the following sections are not "good enough." As a Data Head, you'll need to be an expert in the pitfalls that follow:

- Misapplication of the problem
- Data leakage
- Not splitting your data
- Choosing the right decision threshold
- Misunderstanding accuracy

In particular, misunderstanding accuracy requires its own section, which we'll review after this.

---

[11] See "Ideas on interpreting machine learning" for a good overview: www.oreilly .com/radar/ideas-on-interpreting-machine-learning. There is ongoing research on how to make sense of these methods.

Misapplication of the Problem

It seems obvious, but if you're trying to predict a categorical variable, you shouldn't use linear regression. For instance, recall in Table 10.1 you changed the "yes" and "no" to 1s and 0s to set the problem up for logistic regression.

Your statistical software will not correct you if you wrongly apply linear regression to this data. It doesn't know that your 1s and 0s may mean "Yes" and "No." We've seen it happen many times. So, Data Heads, be advised and kindly correct the mistake when you see it.

Data Leakage

What if, in a rush to build a classification model for the internship offers, you grabbed every piece of historical data possible on intern applications, including whether they were eventually hired or not (0 for "No" and 1 for "Yes"). You then apply logistic regression to predict whether an offer is made.

Can you think of anything wrong with using the attribute *hired*?

Hired indicates if the applicant accepted a full-time position *after* their internship (here's the leak). Only applicants who receive internship offers (which is what you are trying to predict) will have the input "Hired?" = 1. So, if "Hired?" = 1, the target "Offer?" must also always equal 1. Your model is useless because it was trained on data that would not be available at prediction time.

Critical thinking about your data and the input features in a supervised learning algorithm is something software cannot do for your data workers.

Not Splitting Your Data

If you don't split your data into a training set and test set, you risk overfitting the data you have and having terrible performance when new data comes in. By convention, it's recommended to train and learn from 80% of the observations in a dataset and test a model's performance on the other 20%.

Yann LeCun, Chief AI Scientist at Facebook, put it this way: "The act of 'testing on the training set' is anathema in machine learning, the greatest sin you can possibly commit."[12] So make sure you test your models on data it hasn't seen before. If your machine learning algorithm demonstrates near perfect predictions—which is possible gradient boosted trees—your model has likely overfit your training data.

---

[12] Quote taken from a Facebook post: www.facebook.com/ylecun/posts/cmu-statistician-cosma-shazili-received-a-grant-from-the-institute-of-new-econom/318206501719799. Accessed on 9/27/20.

Choosing the Right Decision Threshold

Most classification models don't output a label—they output a probability of belonging to the positive class. Remember how a 2.0 GPA student had a 4% chance of getting an offer while a 3.0 GPA student had a 41% chance? This isn't actionable until a decision rule is introduced.

This is where you come in. The choice of a cutoff probability to determine a final classification is a human decision, not a machine's. Many software packages will choose 0.5, or 50%, as the cutoff as the default. But don't assume this default is representative of your problem.

Do not take this decision cutoff lightly. A model that predicts if someone should receive a credit card offer in the mail might have a low cutoff (this seems to be the case, judging by our mailboxes), while a model that predicts if someone should receive expensive medical treatment might be high. These are tradeoffs you must consider that are heavily reliant on the specifics of your business problem.

Now let's talk about accuracy in classification, and what it even means to be accurate.

MISUNDERSTANDING ACCURACY

As you, and others at your company, are tasked to build, deploy, and defend classification models to automate decisions, you must know how to evaluate and judge these models.

Your first job is to pause and take stock of historical data. As you start to deploy your model, you'll need a yardstick against which to test it. This is called establishing a "control." And you should do it for any classification model you or your data workers build. In binary classification, this is easy: Simply identify the proportion of the *majority class* in your dataset. In the intern applicant dataset, the majority class was "No," as 60% of applicants did not receive offers (and 40% did).

Now let's say someone on your team applies XGBoost (a type of gradient boosted trees algorithm) on 80% of the data (training set), and the classification model predicts correct results 60% of the time on the remaining 20% (test set). That might sound good because it's better than 50/50—you might think, well any information better than a coin flip has a winning expected value over the long term.

In fact, however, it indicates the features in your dataset have no relationship to the output. How can you tell? Well, if you went to your original dataset, completely ignored the inputs, and simply guessed the majority class for each prediction ("No")—you'd be right 60% of the time! As a result, XGBoost didn't help you. The 60% accuracy metric, in a sense, is inaccurate because it doesn't do better than the control.

Consider events that happen infrequently. For instance, an online ad might have thousands of impressions, but only a few people will click it. We would describe this data as being *imbalanced* because the training data is heavily composed of one class (most are "did not click" vs. "click"). If, for instance, 99.5% of people don't click the ad, then simply defaulting to the prediction that no one will ever click the ad will be correct 99.5% of the time.

Because of this, you should not measure machine learning algorithm performance by accuracy alone. There are other, better ways to gauge a classification model's performance using a confusion matrix.

Confusion Matrices

A confusion matrix is a way to visualize the results from a classification model *and* a specific decision threshold. Imagine a random forest model trained on 80% (240 applicants) in the HR intern dataset. The model is then tested on the remaining 20% (60 applicants) to mimic how the model would be used in the real world. The confusion matrix in Table 10.3 shows these results using the default 0.5 cutoff value. Notice that all of the values add up to 60—the number of observations in the test set. In the test set, 23 applicants received an offer and 37 did not. How well did the algorithm do classifying this data?

The confusion matrix gives you several options to assess model performance. Plain accuracy is just one.

Accuracy = Percentage correct = (36 + 19)/60 = 91.6%

But accuracy is rarely what you care about, especially given its vulnerability to imbalanced data. In most applications, you'd probably want to know how well your algorithm predicted true positives and true negatives. In other

TABLE 10.3 Confusion Matrix for Predictions from a Classification Model with a 0.5 Cutoff

		Actuals	
		Yes	No
Predicted	Yes	19	1
	No	4	36

words, is the classifier finding the cases you want it to find (true positives)? Is it ignoring the observations it should ignore (true negatives)?

True Positive Rate (aka "Sensitivity" or "Recall") = Percent of applicants extended an offer divided by how many applicants should have had offers = 19/(19 + 4) = 83%. You want this to be close to 100% if possible.

True Negative Rate ("Specificity") = Percentage of applicants denied an interview divided by how many should have been denied = 36 / (36 + 1) = 97%. You want this to be close to 100% if possible.

Recall previously, the default cutoff to generate a confusion matrix is often 0.5. If we were to increase the cutoff to 0.75, there's now a higher bar for an applicant to receive an offer, and it would change the confusion matrix. The new matrix is shown in Table 10.4.

Notice how the metrics have changed.

True Positive Rate = Percent of applicants extended an offer divided by how many applicants should have had offers = 12/(12 + 11) = 52%.

True Negative Rate = Percentage of applicants denied an interview divided by how many should have been denied = 37 / 37 = 100%.

Increasing the cutoff decreased the true positive rate, which in turn increased the true negative rate. A higher cutoff is perfect at denying applicants who should not be accepted, but this comes at a cost; it also denies several applicants who should have received an offer.

We wanted to demonstrate the tradeoff between the metrics when defining the cutoff. Ultimately, the correct cutoff requires domain expertise to set. As a Data Head, you should spend time considering the best cutoff for your problem.

TABLE 10.4 Confusion Matrix for Predictions from a Classification Model with a 0.75 Cutoff

| | | Actuals | |
		Yes	No
Predicted	Yes	12	0
	No	11	37

Confusing Terms for Confusion Matrices

True positive rate and true negative rate are just a few metrics that can be derived easily from a confusion matrix.

Statisticians and medical doctors call the true positive rate "sensitivity," while data scientists and machine learning engineers might call it "recall." Different fields use different terms for the same metrics.

CHAPTER SUMMARY

In this chapter we presented logistic regression, decision trees, and ensemble methods. We also described many of the pitfalls you'll experience working with classification models. Specifically, we noted common classification pitfalls:

- Misapplication of the problem
- Data leakage
- Not splitting your data
- Choosing the right decision threshold
- Misunderstanding accuracy

In particular, when it comes to understanding accuracy, we described the confusion matrix and how it can be used to better understand model performance. In the next chapter, we'll move to the world of unstructured data to understand text analytics.

CHAPTER 11

Understand Text Analytics

"Seek success but prepare for vegetables."

InspireBot™, an AI bot "dedicated to generating unlimited amounts of unique inspirational quotes."[1]

The last several chapters have dealt with data as we commonly understand it. For most of us, datasets are tables with rows and columns. That's *structured data*. In reality, though, most of the data you interact with every day is *unstructured*. It's in the text you read. It's in the words and sentences of emails, news articles, social media posts, Amazon product reviews, Wikipedia articles, and this book in your hands.

That unstructured textual data is ripe for analysis, but it has to be treated a bit differently. That's what this chapter is about.

EXPECTATIONS OF TEXT ANALYTICS

Before we dive in, we want to set your expectations. Text analytics has received a lot of attention and focus over the years. One example is sentiment analysis—that is the ability to identify the positive or negative emotions behind a

---

[1] Generate your own inspiring quotes at inspirobot.me

social media post, a comment, or a complaint. But as you'll see, text analytics is not such an easy thing to do. By the end of this chapter, you'll have a sense of why some companies can succeed in its use while others will have their work ahead of them.

Many people have preconceived notions about what's possible with computers and human language, undoubtedly influenced by the tremendous success of IBM's Watson computer on the quiz show *Jeopardy!* in 2011[2] and the more recent advancements in speech-recognition systems (think Amazon's Alexa, Apple's Siri, and Google's Assistant). Translations systems, like Google Translate, have achieved near human-level performance by leveraging machine learning (specifically, supervised learning). These applications are held up, rightfully so, as some of the most towering achievements in computer science, linguistics, and machine learning.

Which is why, in our estimation, businesses have exceedingly high expectations when they start analyzing their own text data: customer comments, survey results, medical records—whatever text is stored in your databases. If world travelers can translate their native tongue into one of over 100 languages in a split second, surely it's possible to sift through thousands of your business's customer comments and identify the most pressing issues to your company. Right?

Well, maybe.

Text analytic technologies, while able to solve huge, difficult problems like voice-to-text and speech translation often fail at tasks that seem much easier. And in our experience, when companies analyze their own text data, there's often disappointment and frustration with the results. Put simply, text analytics is harder than you might realize. So, as a Data Head, you'll want to set expectations accordingly.

Our goal in this chapter, then, is to teach you some of the basics of text analytics,[3] which is extracting useful insights from raw text. To be clear, we can only scratch the surface of this burgeoning field, but we hope to provide you with enough information to give you a feel of the possibilities and challenges of text analytics. As new developments come about in this space, you'll have the tools to understand what will help and what won't. And, as with any topic, the more you learn, the more you'll naturally build awareness to what's possible but also come away with some warranted Data Head skepticism.

In the sections that follow, we'll talk about how to gain structure from your unstructured text data, what kind of analysis you can do on it, and then

---

[2] There's a great description of Watson's question-answering system in the book: Siegel, E. (2013). *Predictive analytics: The power to predict who will click, buy, lie, or die.* John Wiley & Sons.

[3] You might also hear *text mining*.

we'll revisit why Big Tech can make seemingly science fiction-like progress with their text data while the rest of us might struggle.

HOW TEXT BECOMES NUMBERS

When humans read text, we see mood, sarcasm, innuendo, nuance, and meaning. Sometimes that feeling is even unexplainable: a poem is evocative of a memory; a joke makes you laugh.

So it's likely not a surprise that a computer does not understand meaning the way a human does. Computers can only "see" and "read" numbers. The mass of unstructured text data must first be converted to numbers and the structured datasets you're familiar with in order to be analyzed. This process—converting unstructured and potentially messy text with misspellings, slang, emojis, or acronyms into a tidy, structured dataset with rows and columns—can be a subjective and time-consuming process for you and your data workers. There are several ways to do it, and we'll cover three.

A Big Bag of Words

The most basic approach to convert text into numbers is by creating a "bag of words" model. In a bag of words, a sentence of text is jumbled together into a "bag" where word order and grammar are ignored. What you read as: "This sentence is a big bag of words" is converted into a set, called a *document*, where each word is an identifier, and the count of the word is a feature. Order does not matter, so we'll sort the bag alphabetically by identifier:

{a: 1, bag: 1, big: 1, is: 1, of: 1, sentence: 1, this: 1, words: 1}.

Each identifier is referred to as a *token*. The entire set of tokens from all documents is called a *dictionary*.

Of course, your text data will contain more than one *document*, so this bag of words can get exceedingly large. Every unique word and spelling would become a new token. Here's how this would look as a table, where each row contains a sentence (or customer comment, product review, etc.).

For the raw text:

- This sentence is a big bag of words.
- This is a big bag of groceries.
- Your sentence is two years.

The bag-of-words representation would look like that in Table 11.1, where the data points represent the count of each word in the sentence.

TABLE 11.1 Converting Text to Numbers as a Bag of Words. The numbers represent how many times each word (token) appears in the corresponding sentence (document).

Original Text	a	bag	big	groceries	is	of	sentence	this	two	words	years	your
This sentence is a big bag of words.	1	1	1	0	1	1	1	1	0	1	0	0
This is a big bag of groceries.	1	1	1	1	1	1	0	1	0	0	0	0
Your sentence is two years.	0	0	0	0	1	0	1	0	1	0	1	1

Looking at Table 11.1, called a *document-term matrix* (one document per row, one term per column), you'll notice how easy it would be to perform some basic text analytics: calculating summary statistics of each word (*"is"* is the most popular word) and finding out which sentence contains the most tokens (first sentence). While not interesting in this example, this is how basic summary statistics of documents can be calculated.

Quick Thoughts on Word Clouds

Before we move forward, let's talk about word clouds. Word clouds are often people's first experience with text analytics. It's a simple visual where the size of the word in the "cloud" indicates how often it appears in the dictionary. The word cloud for this chapter is shown in Figure 11.1.[4]

FIGURE 11.1   A word cloud for the text in this chapter

Did you learn anything useful from Figure 11.1? Probably not. We understand that word clouds make great marketing material, but we're not fans and we do not recommend them. Even as a visualization tool, it's harder to interpret size in area than a simple length, where each word could be represented as a category in a bar chart, with the bar's length showing the frequency of the word.

---

[4] Word cloud generated by wordclouds.com.

We suspect you've also noticed some drawbacks in Table 11.1 (and hopefully word clouds!). As more documents are added, the number of columns in the table would become exceedingly wide because you'd have to add a new column for each new token. The table would also become very *sparse*—full of zeros—because each individual sentence would only contain a handful of words from the dictionary.

To combat this, it's standard practice to remove common filler words— like *the, of, a, is, an, this*, etc.—that don't add meaning or differentiation between sentences. These are called *stop words*. It's also common to remove punctuation and numbers, convert everything to lowercase, and *stem* words (cut off their endings) to map words like *grocery* and *groceries* to the same stem *groceri*, or *reading, read, reads* to *read*. A more advanced counterpart to stemming is called *lemmatization*, which would map words *good, better, best* to the root word *good*. In that sense, lemmatization is "smarter" than stemming, but it takes longer to process.

Little adjustments like this can drastically reduce the size of the dictionary and make analysis easier. Figure 11.2 shows what this process would look like for one sentence.

It should come as no surprise then, after seeing this approach in Figure 11.2, why analyzing text is hard. The process to convert text into numbers has filtered out emotion, context, and word order. If you suspect this would impact the results of any subsequent analysis, you'd be right. And we were lucky here—there aren't any misspellings, which are an added challenge for data workers.

Using the bag-of-words approach, which is available in free software and the first approach data workers learn in text analytics courses, would provide

Text To Numbers	Text Processing Steps
You are reading a short, simple sentence with 10 words!	Convert to lowercase, remove punctuation
you are reading a short simple sentence with 10 words	Remove stop words and numbers
reading short simple sentence words	Stem the words
read short simpl sentenc word	Count each token
read: 1, sentenc: 1, short:1, simpl:1, word:1	Final Result

FIGURE 11.2   Processing text down to a bag of words

the same numerical encoding for the following two sentences, despite the obvious difference in meaning:

1. Jordan loves hotdogs but hates hamburgers.[5]
2. Jordan hates hotdogs but loves hamburgers.

Humans know the difference between those two sentences. A bag-of-words approach does not. But make no mistake. Despite its simplistic approach, a bag of words can be helpful when summarizing very disparate topics, which we'll highlight in later sections.

N-Grams

It's easy to see where the bag-of-words approach let us down in the example about Jordan and his hotdogs. The two-word phrase "loves hotdogs" is the opposite of "hates hotdogs," but bag-of-words throws context and word order aside. N-grams can help. An *N-gram* is a sequence of N consecutive words, so the 2-grams (formally called *bigrams*) for the sentence "Jordan loves hotdogs but hates hamburgers" would be: {but hates: 1, hates hamburgers: 1, hotdogs but: 1, Jordan loves: 1, loves hotdogs: 1}

It's an extension of bag-of-words but adds the context we need to differentiate phrases with identical words in different arrangements. It's common to add the bigram tokens to the bag-of-words, which further increases the size and sparseness of the document-term matrix. From a practical matter, this means you need to store a large (wide) table with relatively little information inside. We added some bigrams to Table 11.1 to create Table 11.2.

There are some debates whether to filter out bigrams with stop words, as context can be lost. The words "my" and "your" are considered stop words in some software, but the meaning behind the bigram phrases "my preference" and "your preference" is gone if stop words are removed. This is another decision your data workers must make when analyzing text data. (Can you start to sense why text analytics is challenging?)

But once prepared, simple counts can be helpful to summarize text. Websites like Tripadvisor.com take advantage of these approaches and give users the ability to quickly search reviews for frequently mentioned words or

---

[5] Jordan's favorite food is hotdogs.

TABLE 11.2 Extending the Bag-of-Words Table with Bigrams. The resulting document-term matrix gets very wide.

a	bag	big	groceries	is	of	sentence	this	two	words	years	your	a big	big bag	is a	sentence is	this sentence	of groceries	this is	bag of	
1	1	1	0	1	1	1	1	0	1	0	0	1	1	1	1	1	0	0	1	…
1	1	1	1	1	1	0	1	0	0	0	0	1	1	1	0	0	1	1	1	…
0	0	0	0	1	0	1	0	1	0	1	1	0	0	0	1	0	0	0	0	…

phrases. You might see, for example, suggested searches like "baked potato" or "prepared perfectly" as commonly mentioned bigrams at your local steakhouse.

Word Embeddings

With bag-of-words and N-grams, it's possible to tell similarity between documents: if multiple documents contain similar sets of words or N-grams, you can reasonably assume the sentences are related (within reason, of course. Don't forget about Jordan's love/hate relationship with hotdogs). The rows in the document-term matrix would be numerically similar.

But how could someone go about discovering, numerically, which words in a dictionary—not documents, but words—are related?

In 2013, Google analyzed billions of word pairs (two words within close proximity of each other within a sentence) in its enormous database of Google News articles.[6] By analyzing how often word pairs appeared—for example, *(delicious, beef)* and *(delicious, pork)* appeared more than *(delicious, cow)* and *(delicious, pig)*—they were able to generate what are called *word embeddings*, which are numeric representations of words as a list of numbers (aka vectors). If *beef* and *pork* appear often with the word *delicious*, the math would represent each word as being similar in an element of the vector associated with things described as delicious, or what we humans would simply call food.

To explain how it works, we'll use a tiny set of word pairs (Google used billions). Suppose we scan a local newspaper article and find the following

---

[6] A more in-depth description of Word2vec is presented in Chapter 11 of the outstanding book: Mitchell, M. (2019). *Artificial intelligence: A guide for thinking humans*. Penguin UK.

pairs of words: *(delicious, beef), (delicious, salad), (feed, cow), (beef, cow), (pig, pork), (pork, salad), (salad, beef), (eat, pork), (cow, farm)*, etc. (Imagine several more pairs along this line of thinking.) Every word goes into the dictionary: {*beef, cow, delicious, farm, feed, pig, pork, salad*}.[7]

The word *cow*, for example, can be represented as a vector the length of the preceding dictionary—one component per word with a 1 in the location of *cow* and zero otherwise: (0, 1, 0, 0, 0, 0, 0, 0). This is an input into a supervised learning algorithm that is mapped to its associated output vector (also the length of the dictionary). In it are the probabilities that the other words in the dictionary appeared near the input word. So for input *cow*, the associated output might be (0.3, 0, 0, 0.5, 0.1, 0.1, 0, 0), to show *cow* was paired with *beef* 30% of the time, *farm* 50% of the time, and *feed* & *pig* 10% of the time.

As is the goal with every supervised learning problem, the model tries to map the inputs (word vectors) as close as possible to the outputs (vectors with probabilities). But here's the twist. We don't care about the model itself; we care about a piece of math generated from the model: a table of numbers that shows how each word in the dictionary relates to every other word. These are the word embeddings. Think of them as a numeric representation of a word that encodes its "meaning." Table 11.3 shows some of the words in our dictionary, along with their word embeddings across the rows. *Cow*, for example, is written as a three-dimensional vector (1.0, 0.1, 1.0). Before, it was written as the longer, sparser vector (0, 1, 0, 0, 0, 0, 0, 0).

What's fascinating about word embeddings is that the dimensions (hopefully) contain meaning behind the words, similar in a way to how the reduced dimensions in PCA captured themes of features.

TABLE 11.3 Representing Words as Vectors with Word Embeddings

Word	Dimension 1	Dimension 2	Dimension 3
Cow	**1.0**	0.1	**1.0**
Beef	0.1	**1.0**	**0.9**
Pig	**1.0**	0.1	0.0
Pork	0.1	**1.0**	0.0
Salad	0.0	**1.0**	0.0

---

[7]Yes, we are ignoring many pairs of words that would occur, even in the shortest of articles. This alone should convince you of the computational challenge Google had to undertake to do this.

Look down Dimension 1 in Table 11.3. Can you spot a pattern? Whatever Dimension 1 means, *Cow* and *Pig* have a lot of it, and *Salad* has none of it. We might choose to call this Dimension *Animal*. Dimension 2, we can call *Food* because *Beef*, *Pork*, and *Salad* scored high, and Dimension 3 *Bovine* because words associated with cows stood out. With this structure, it's now possible to see similarities in how words are used. It's even possible (if a bit weird) to show simple equations between words.

As an exercise, convince yourself (using Table 11.3) that *Beef – Cow + Pig ≈ Pork*.[8]

This technique is called Word2vec[9] (word to vector) and the word embeddings Google generated for us are freely available to download.[10] Of course, don't expect every relationship to be perfect. There's variation in all things, as you're astutely aware of by now, and it's certainly present in text. Non-food items can be described as delicious, like *delicious irony*. There's *orange* the color and *orange* the juice, and a plethora of homonyms in our language to complicate things.

Word embeddings, with their numeric structure that enables calculation, have applications in search engines and recommendation systems, but the word embeddings generated from Google News text might not be specific to your problem set. For example, brand names like Tide® (laundry detergent) and Goldfish® (crackers) may be semantically similar to the words "ocean" and "pet," respectively, in a system like Word2vec, but for a grocery store, those items would be semantically more similar to competing brands like Gain® detergent and Barnum's Animals® crackers.

It is possible to run Word2vec on your text data and generate your own word embeddings. This would help you glean topics and concepts you might otherwise not even consider in your data. However, getting enough data to be meaningful is an issue. Not every company will have access to as much data as Google. You simply may not have enough text data to find meaningful word embeddings.

TOPIC MODELING

Once we have our text into a meaningful dataset, we're ready to start some analysis. And, there's a satisfying payoff to this investment of converting unstructured text data into a structured dataset with rows and columns of

---

[8] *Beef* = (0.1, 1, 0.9), *Cow* = (1.0, .1, 1.0), *Pig* = (1.0, 0.1, 0.0). Add/subtract the elements. *Beef – Cow + Pig* = (0.1, 1, -0.1), which is close to *Pork* = (0.1, 1.0, 0).
[9] Mikolov, T., Chen, K., Corrado, G., & Dean, J. (2013). Efficient estimation of word representations in vector space. arXiv preprint arXiv:1301.3781.
[10] code.google.com/archive/p/word2vec

numbers. That's because we can use those analysis methods you've been learning about throughout this book (with some tweaks). Let's talk about those analysis methods in the next few sections.

In Chapter 8, you learned about unsupervised learning—finding natural patterns in the rows and columns of a dataset. Unleashing a clustering algorithm like k-means on a document-term matrix, like in Tables 11.1 and 11.2, would produce a set of k distinct groups of text that are similar in some way. This could prove to be helpful in some cases. But applying k-means clustering to text data is quite rigid. Consider, for instance, these three sentences:

1. The Department of Defense should outline an official policy for outer space.
2. The Treaty on the Non-Proliferation of Nuclear Weapons is important for national defense
3. The United States space program recently sent two astronauts into space.

In our eyes, there are two general topics being discussed here: national defense and space. The first sentence discusses both topics, while the second and third touch on one. (If you disagree, and you're certainly allowed to, then you are realizing a central challenge of clustering text together—there's not always a clean separation of topics. And with text, unlike a row of numbers, everyone can quickly form an opinion.)

Topic modeling[11] is similar in spirit to k-means in the sense that it's an unsupervised learning algorithm that attempts to group similar observations together, but it relaxes the notion that each document should be explicitly assigned to one cluster. Instead, it provides probabilities to show how one document might span across several topics. Sentence 1, for example, might score 60% for national defense and 40% for space.

A new example will help. In Figure 11.3, you see a visual representation of a document-term matrix flipped on its side.[12] Along the left, you see the terms that appear in the 20 documents along the top, denoted d0 though d19. Each cell represents the frequency of a word in the document, with darker cells indicating a higher frequency. But the terms and documents have been arranged, via topic modeling, to reveal the results of the topic model.

Study the image, and you'll spot which words appear frequently in documents together forming possible topics, as well as documents that contain

---

[11] Two popular types of topic modeling are Latent Semantic Analysis (LSA) and Latent Dirichlet Allocation (LDA).

[12] This figure is from en.wikipedia.org/wiki/File:Topic_model_scheme.webm, created by Christoph Carl King and made available on Wikipedia under the Creative Commons Attribution-Share Alike 4.0 International license.

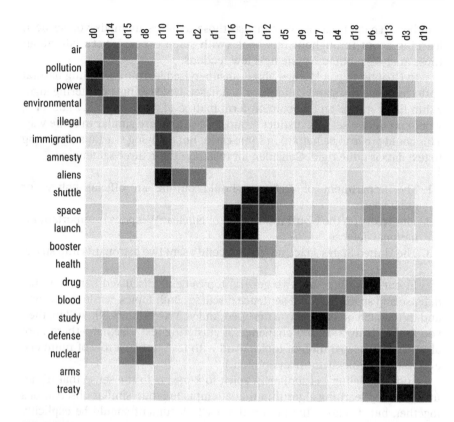

FIGURE 11.3 Clustering documents and terms together with topic modeling. Can you spot the five main topics in the image? What would you name them?

several terms that are woven across multiple topics (look at d13 specifically, the third column from the right). Be advised, though—like all unsupervised learning methods, there's no guarantee of accurate results.

As a practical matter, topic modeling works best when there are disparate topics within your set of documents. That may seem obvious, but we've seen instances where topic modeling has been applied on subsets of text that have been filtered down to a specific topic of interest prior to the analysis. This would be like taking a group of news articles, searching for only those that contain "basketball" and "LeBron James" and then expecting topic modeling to break out the remaining articles into something meaningful. You'll be disappointed with the results. By filtering the text, you've effectively defined one topic for the articles that remain. Keep this nuance in mind, continue to argue with your data, and manage expectations appropriately.

TEXT CLASSIFICATION

In this section, we'll switch from unsupervised learning to supervised learning on a document-term matrix (provided a known target exists to learn from). With text, we're usually trying to predict a categorical variable, so this falls under the classification models you learned last chapter, as opposed to regression models that predict numbers. One of the most well-known text classification success stories is the spam filter in your email, where the input is the text of an email, and the output is a binary flag of "spam" or "not spam."[13] A multiclass classification application of text analytics is the automatic assignment of online news articles to the news categories: Local, Politics, World, Sports, Entertainment, etc.

Let's look at one (oversimplified) case to give you an idea how text classification works using bag-of-words. Table 11.4 shows five different email subject lines, broken out into tokens, and a label indicating if the email is spam or not. (As an aside, let's not discount the effort companies take to collect data like this. There's a reason your email provider asks you if your email is spam or not. You are providing the training data for machine learning algorithms!)

How might we use an algorithm to learn from the data in Table 11.4 to make predictions about new, unseen email subject lines?

TABLE 11.4 A Basic Spam Classifier Example

Email Subject	advice	bald	birthday	debt	free	help	mom	party	relief	stock	viagra	Spam?
Advice for Mom's birthday party	1	0	1	0	0	0	1	1	0	0	0	0
Free Viagra!	0	0	0	0	1	0	0	0	0	0	1	1
Free Stock Advice	1	0	0	0	1	0	0	0	0	1	0	1
Free Debt Relief Advice	1	0	0	1	1	0	0	0	1	0	0	1
Balding? We can help!	0	1	0	0	0	1	0	0	0	0	0	1

---

[13] Drucker, H., Wu, D., & Vapnik, V. N. (1999). Support vector machines for spam categorization. *IEEE Transactions on Neural networks*, *10*(5), 1048–1054 is one of the seminal papers in this area.

Perhaps you thought of logistic regression, which is useful to predict binary outcomes. Unfortunately, logistic regression will not work here. Why not? The math behind logistic regression breaks down because there are too many words and not enough examples to learn from. There are more columns than rows in Table 11.4, and logistic regression doesn't like that.[14]

Naïve Bayes

The popular method to apply in this situation is a classification algorithm called Naïve Bayes (named after the very same Bayes we mentioned in Chapter 6). The intuition behind it is straightforward: are the words in the email subject line more likely to appear in a spam email or a non-spam email? You probably perform a similar process to this when you read your inbox: the word "free," based on your experience, is typically a spammy word. As is "money," "Viagra," or "rich." If most words are spammy, the email is probably spam. Easy.

Said another way, you're trying to calculate the probability an email is spam, given the words in the subject line, $(w_1, w_2, w_3, \ldots)$. If this probability is greater than the probability it's not spam, mark the email as spam. In probability notation, these competing probabilities are written as:

- Probability an email is spam = $P(\text{spam} \mid w_1, w_2, w_3, \ldots)$
- Probability an email is not spam = $P(\text{not spam} \mid w_1, w_2, w_3, \ldots)$

Before moving on, let's take stock of the data available to us in Table 11.4. We have the probability that each word occurs in spam (and not spam) emails. "Free" occurred in three out of the four spam emails, so the probability of seeing "free," given the email is spam, is $P(\text{free} \mid \text{spam}) = 0.75$. Similar calculations would show: $P(\text{debt} \mid \text{spam}) = 0.25$, $P(\text{mom} \mid \text{not spam}) = 1$, and so on.

What does that give us? We want to know the probability an email is spam given the words, but what we have is the probability of seeing a word, given it's spam. These two probabilities are not the same, but they are linked through Bayes' Theorem (from Chapter 6). Recall the central idea behind Bayes' is to swap the conditional probabilities. Thus, instead of working with $P(\text{spam} \mid w_1, w_2, w_3, \ldots)$, we can use $P(w_1, w_2, w_3, \ldots \mid \text{spam})$. Through some

---

[14] Linear regression also does not work if there are more features than observations in the data. However, there are variants of linear and logistic regression that can handle more features than observations.

additional math (which we're skipping for brevity[15]), the decision to classify a new email as spam comes down to figuring out which value is higher:

1. Spam score = P(spam) × $P(w_1 \mid$ spam) × $P(w_2 \mid$ spam) × $P(w_3 \mid$ spam)
2. Not Spam score = P(not spam) × $P(w_1 \mid$ not spam) × $P(w_2 \mid$ not spam) × $P(w_3 \mid$ not spam)

All of this information is available in Table 11.4. The probabilities P(spam) and P(not spam) represent the proportion of spam and not spam in the training data—80% and 20%, respectively. In other words, if you had to take a guess without looking at the subject line, you'd guess "spam" because that's the majority class in the training data.

In order to get to the preceding formulas, the Naïve Bayes approach committed what is usually an egregious error in probability—assuming independence between events. The probability of "free" and "Viagra" appearing in a spam email together, denoted P(free, viagra | spam) depends on how often both words appear in the same email, but this makes the computation much more difficult. The "naïve" part of Naïve Bayes is to assume every probability is independent: P(free, viagra | spam) = P(free | spam) × P(viagra | spam).

A Deeper Look

If you see an email with the subject line "Get out of debt with our stock advice!" you'd focus on the non-stop words *get, debt, stock,* and *advice* and calculate the competing values:

1. Spam score = $P\left(\text{spam}\right) \times P\left(\text{get} \mid \text{spam}\right) \times P\left(\text{debt} \mid \text{spam}\right) \times P\left(\text{stock} \mid \text{spam}\right) \times P\left(\text{advice} \mid \text{spam}\right)$

2. Not Spam score = $P\left(\text{not spam}\right) \times P\left(\text{get} \mid \text{not spam}\right) \times P\left(\text{debt} \mid \text{not spam}\right) \times P\left(\text{stock} \mid \text{not spam}\right) \times P\left(\text{advice} \mid \text{not spam}\right)$

---

[15] For more details, see the Wikipedia article on "Naïve Bayes spam filtering."

But there's a slight problem. New and rare words require some adjustments to calculations to avoid multiplying the probabilities by zero. In the tiny dataset in Table 11.4, the word "get" doesn't appear at all, and the words debt, stock, and advice have only shown up in spam. These quirks would make both the spam and not spam scores equal to zero. To fix this, let's pretend we've seen each word at least once by adding 1 to the frequency of each word. We'll also add 2 to the frequency of the spam (and not spam) to prevent values from reaching 1.[16]

Now we can calculate:

1. Spam Score: $(0.8) \times \dfrac{0+1}{4+2} \times \dfrac{1+1}{4+2} \times \dfrac{1+1}{4+2} \times \dfrac{2+1}{4+2} = 0.0074$

2. Not Spam Score: $(0.2) \times \dfrac{0+1}{1+2} \times \dfrac{0+1}{1+2} \times \dfrac{0+1}{1+2} \times \dfrac{1+1}{1+2} = 0.0049$

The top number is larger, so we'd predict spam for the email: "Get out of debt with our stock advice!"

Sentiment Analysis

Sentiment analysis is a popular text classification application on social media data. If you search Google for "sentiment analysis of Twitter data," you might be surprised by the number of results; it seems everyone is doing it. The underlying idea is like the spam/not spam example earlier: are the words in a social media post (or product review, or survey) more likely to be "positive" or "negative." What you do with this information depends on your business cases, but there is an important callout to mention with sentiment analysis: do not extrapolate beyond the context of the training data and expect meaningful results.

What do we mean? Many "sentiment analysis" classifiers learn from freely available data online. A popular dataset for students is a large collection of movie reviewers from the Internet Movie Database (IMDb.com). This collection of data, and any model you build on it, would be relevant to movie reviews only. Sure, it would associate words like "great" and "awesome" with positive sentiment, but don't expect it to perform well if you have a unique business case with its own vernacular.

---

[16] This is called a Laplace correction. It helps prevent high variability caused by low counts, something we discussed in Chapter 3.

What About Tree-Based Methods on Text?

Tree-based methods like random forests and boosting can be applied to text classification problems and tend to perform better than Naïve Bayes on some datasets, but Naïve Bayes is usually a good start and has a transparent interpretation.

PRACTICAL CONSIDERATIONS WHEN WORKING WITH TEXT

Now that you're familiar with a handful of tools in the text analytics toolbox, we're going to take a step back in this section and talk about text analytics at a high level.

When working with text, you have the luxury of reading the data. If topic modeling suggests certain sentences belong in topics, review the results for a gut check. If someone builds a text classification model, ask to see the results: the good, the bad, and the ugly.

From experience, a successful text analytics project is fun to present to stakeholders because the audience can read the data and participate in conversations about the results—it's not a series of numbers, but something they can read, understand, and pass judgment of immediately. But presenters are tempted to show the exciting examples and the easy wins, rather than the clear misses. Data Heads, if presenting text analytic results, should be transparent with outcomes. Likewise, if you are consuming results, request to see examples where the algorithms went wrong. Trust us, they exist.

Which brings us back to a comment we made early in the chapter: *when companies analyze their own text data, there's often disappointment and frustration with the results.* This wasn't to turn you away from text—far from it. By being transparent with the shortcomings, we hope we can prevent a possible backlash where you or your company starts analyzing text, realizes it's trickier than anticipated, and angrily dismisses it entirely or ignores its lack of usefulness and goes full steam ahead with a weak analysis.

By now, you've developed enough skepticism in the earlier sections to realize where hiccups can emerge. But some big tech companies have seemed to rise above those challenges and have emerged as leaders in text analytics and *Natural Language Processing* (NLP), which deals with all aspects of language, including audio (as opposed to just written text).

Big Tech Has the Upper Hand

Here's what big tech companies like Facebook, Apple, Amazon, Google, and Microsoft have that many other companies don't have: an abundance of text and voice data (an abundance of *labeled* data that can be used to train supervised learning models); powerful computers; dedicated (and world-class) research teams; and money.

With these resources, they've made remarkable progress not only with text, but also with audio. In recent years, there has been noticeable improvements to the areas of

- **Speech-to-Text**: Voice-activated assistants and voice-to-text on smartphones are more accurate.
- **Text-to-Speech**: Computers' read-aloud voices sound more human-like.
- **Text-to-Text**: Translating one language to another is instantaneous with good accuracy.
- **Chatbots**: The automated chat windows that pop up on every website now with: "How may I help you?" are (somewhat) more helpful.
- **Generating Human-Readable Text**: The language model known as GPT-3[17] from OpenAI is a language model that can generate human-like text (you might think a human wrote it), answer questions, and generate computer code on demand. And it is, as of this writing, the most advanced model of its kind. Estimates put the cost of training the model (not paying the researchers, just running the computers) at $4.6 million.[18]

Add to that the access to data and an expert research team, and you get the idea how NLP is often a case of the haves vs. the have nots. Most companies, to be blunt, aren't there (yet?). Even though the algorithms are open source, the massive collection of data and access to supercomputers is not. Big tech clearly has the advantage.

The other thing to consider, when establishing expectations, is how many of the applications from big tech are universal to millions of people; think of them as general tasks common to all sects of society. Amazon Alexa is designed to work for everyone, including children. And when translating text, there are rigid rules built into the training data. The word *party* in English is the word *fiesta* in Spanish. Our point is this: everyone who uses these systems is expecting it to work in the same way.

---

[17] Generative Pre-trained Transformer 3
[18] www.forbes.com/sites/bernardmarr/2020/10/05/what-is-gpt-3-and-why-is-it-revolutionizing-artificial-intelligence/#116e7b04481a

Contrast this to a business-specific text classification task. For example, the sentiment of the sentence "Samsung is better than an iPhone" depends on whether you work for Apple or Samsung. The data you have access to may have its own type of language, unique to only your company. Not only that, but the size of the data will be smaller than what tech companies have available. Consequently, results may not be as clean as you would expect.

Still, we strongly encourage you to avail yourself of all available algorithms including text analytics. Insight is not a matter of big machines, but rather a matter of context and expectation. If you understand the limitations of text analytics before you start, you'll be prepared to run it correctly at your company.

CHAPTER SUMMARY

By going through this chapter, we hope we've convinced you that computers don't understand language as humans do—to a computer, it's all numbers. This alone, in our opinion, is incredibly valuable to know. You're less likely to be swindled the next time you hear marketing rhetoric about how *artificial intelligence* can solve every text-related business problem you can think of because the process of converting text to numbers removes some of the meaning we as humans would assign to words and sentences. We discussed three methods:

1. Bag of words
2. N-grams
3. Word embeddings

When converted to numbers, you can apply unsupervised learning tasks like topic modeling or supervised learning tasks like text classification. Finally, we described how Big Tech has the upper hand, so set your expectations accordingly based on the amount of data and resources you have.

In the next chapter, we'll continue our analysis of unstructured data to describe neural networks and deep learning.

CHAPTER **12**

Conceptualize Deep Learning

"AI is sometimes heralded as the new industrial revolution. If deep learning is the steam engine of this revolution, then data is its coal: the raw material that powers our intelligent machines, without which nothing would be possible."

—François Chollet, AI researcher and author[1]

Congratulations on making it this far. This chapter, in many ways, serves as the climax of your Data Head journey. It's where we weave several pieces together and lay the groundwork to the growing subfield of machine learning called *deep learning*.

Today, the use of deep learning drives bleeding-edge technologies, and our collective fascination with it is that it can sometimes appear to be all too human. Deep learning comprises the group of technologies that drive facial recognition, autonomous driving, cancer detection, and speech translation. That is, they help drive decisions that were once considered the province of humans. Though, as we'll describe here soon, deep learning is neither new nor as much like the human mind as you might have heard.

That said, much of the excitement, promise, and hype in the data space concerns the potential of deep learning. The business world, unsurprisingly, is

---

[1]Chollet, F. (2018). *Deep learning with Python*. New York: Manning.

investing piles of money to increase deep learning's footprint, and it's poised to impact many industries in the coming years. But as deep learning grows, so does its hype—and often overlooked are the ethical concerns it creates.

In this chapter, we'll reveal the components of deep learning. First, we'll start with its structure. At its core, deep learning uses a family of models called artificial neural networks. These algorithms are said to mimic the way the brain thinks through an idea; but as we shall see, they only kinda sorta do this. Next, we'll go deeper, as it were, and talk about the ways neural networks can be modified to take on more complicated learning tasks (like image recognition). Finally, we'll conclude with the reality of deep learning, focusing on its practical challenges, how often they're incorrectly deployed, and the broader implications of running black box models.

NEURAL NETWORKS

To begin conceptualizing deep learning, it's necessary to first understand artificial neural networks, the building blocks behind deep learning.

How Are Neural Networks Like the Brain?

The human brain is comprised of a network of biological neurons. These neurons are said to "absorb information" in the form of chemical signals and electrical impulses. At a certain point—and not exactly at a point we truly understand—the information causes a neuron to "fire" or react. If you're driving a car and a deer runs in front of you, your brain quickly processes the inputs (your speed, distance to the deer, nearby traffic), causing millions of neurons to fire, which then triggers an output (either slam on the brakes or swerve out of the way).[2]

The question then becomes: can we create a series of models and algorithms that can learn the same way our brain learns? Can we quickly transform inputs in the form of data, images, or sound into a meaningful output? If we could mimic our brain into an algorithm, imagine the possibilities. How many human evaluations which we do second by second could we optimize and have a computer do for us?

Artificial neural networks, the computational counterpart to biological neural networks, were created to try to answer this question.

---

[2]Of course, a deer jumping in front of your car is an extreme example to demonstrate a sharp and expected change in brain chemistry. The reality is that your brain is quickly processing inputs and output rights now. Millions of neurons are currently firing as you read this.

That all sounds incredible, and to be sure, your authors do find neural networks fascinating. The original neural networks were created in the 1940s and were designed to mimic human biology as it was understood at the time. Indeed, much of the hype around neural networks—and thus deep learning—stems from the fact that they're inspired by our brains. But the risk with the "neural network is like the brain" analogy is that it projects a level of abstraction and general knowledge onto neural network models that, in reality, are just giant math equations.

So, despite what you might hear in popular media or sales calls, we should not be fooled into thinking the latest advancements in neural networks and deep learning reflect a closer link to the human brain. Rather, the success of these algorithms stems from faster computers, piles of data, and a wave of research in machine learning, statistics, and mathematics.

Let's now see how neural networks work by walking through two examples.

A Simple Neural Network

Recall from Chapter 10, Understand the Classification Model, that we built a model to predict if an internship applicant should be offered an interview based on their GPA, year in school, major, and number of extracurricular activities. Figure 12.1 depicts how this would be visualized as the most basic neural network.

Figure 12.1 shows four inputs for an applicant:

- GPA = 3.90
- Year in School = 4
- Major = "Statistics" (encoded as a 2)
- Extracurriculars = 5 (reflects the total number of extracurricular activities)

These values flow into the computational unit called a *neuron*, represented as a circle in our figure. Inside this neuron is an *activation function*.

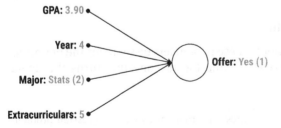

FIGURE 12.1 The simplest neural network possible. The four inputs are processed by an activation function in the single neuron, producing an output.

The activation function converts the four input values into a single numerical output. The biological motivation being, if some combination of the inputs surpasses a threshold, the neuron will "fire" and predict the applicant should get an offer.

Several functions can be used for the activation function, depending on the problem you're trying to solve and the data you have available. Because we are dealing with a classification problem—will this intern receive an offer?—our activation function is designed to produce the probability of an offer, just like we did in Chapter 10 using logistic regression.[3]

In Equations 12.1 and 12.2 we show you a common activation function. We've broken it up into two parts to make it easier to work with (and to print):

$$\text{Probability of Offer} = \frac{1}{1+e^{-(x)}} \tag{12.1}$$

$$\text{Where } X = w_1 * GPA + w_2 * Year + w_3 * Major + w_4 * Extracurriculars + b \tag{12.2}$$

Hopefully these equations feel familiar by now. Equation 12.1 is the logistic function from Chapter 10 and Equation 12.2 is the linear regression function introduced in Chapter 9. So, mathematically, a neural network simply contains components of past machine learning and statistical algorithms. The linear regression-line equation in 12.2 allows us to combine the four inputs into one, and the logistic function in 12.1 squishes the result to be between 0 and 1, the range where probabilities must lie.

The goal of the network, as is the goal of logistic regression, is to find the best values for the weights and constant term (represented as w's and b in 12.2, respectively; collectively called parameters) that make the predicted outputs of the network as close to the actual outputs in the aggregate.[4] The "learning" part of a neural network (and machine learning in general) refers to the training process that optimizes the parameters of equations like 12.2 for prediction.

How a Neural Network Learns

The real question is what should these parameters be set to for the optimum? That's the magic answer we're looking for, the one that turns the neural

---

[3]Neural networks can be used for regression problems as well. Regression problems would have a different activation function as the final calculation—essentially a linear regression model.

[4]Weights are also called coefficients. There are multiple names for the same concepts.

network into a useful prediction machine. But at the start of the training process, the parameters could really be anything. So our algorithm assigns them a random value to start with because you have to start somewhere. If you need to wash your hands and you were at the tap for the first time—and the tap did not have clear indications for hot and cold—you would just turn it on and test the temp. You'd modulate from there. Same thing here.

These starting random weights are inherently wrong: wrong in the sense that they were pulled from random and not *learned*. But they get the ball rolling. And, most importantly, they produce a numeric output. For example, let's say you take two exceptional past intern applicants, call them Will and Allie, and run their input data through the neural network (i.e., the preceding equations). Suppose our randomized parameters produced the output 0.2 for Will and 0.3 for Allie. In other words, the random weight and constant parameters indicated a low probability of an offer for both applicants. But remember, this is historical training data—we know what the output values should have been. And in this case, both Will and Allie received offers. Their true output values were 1, but the model predicted low values for each. So to start, the neural network is a terrible predicter.

At this point, the model looks at the true values of the outputs (1 and 1) and sends a message that the current parameters are wrong; adjust them. But in what direction should we adjust each weight and by how much? An algorithm called backpropagation[5] adjusts these weights and decides to increase or decrease them and by how much: Should GPA matter more? Perhaps Year should matter less? The process then repeats: the updated weights are used to score Will and Allie's data again, this time producing the output 0.4 and 0.6. Better, but not great. Backpropagation sends a signal back through the network and adjusts the weights again. Rinse and repeat. Over time the parameters converge toward their hypothetical optimum that produces the closest predicted values to the actual labels, on average.[6]

A Slightly More Complex Neural Network

In the previous example, all we did was take logistic regression and turn it into a neural network visualization. The math was identical. You may be asking, then, why even do this? Why represent logistic regression as something new called a neural network?

---

[5]For any Calculus fans out there, backpropagation is essentially the chain rule, which provides the tools to optimize nested equations like those we find in neural networks.
[6]In linear regression, there is a true mathematical optimum for the parameters (that is, there is a point at which the sum of squares can no longer be reduced). Unfortunately, for neural networks, there's often no way to know if our neural network has reached a mathematical optimum or just a convenient, "good enough" result.

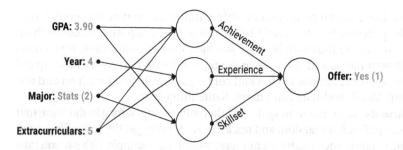

FIGURE 12.2 A neural network with a hidden layer. The middle layer is "hidden" between the main input layer on the left and output layer on the right.

The answer to that question, and the real benefit of neural networks, lies in what happens when you add "hidden" layers into the network. So, let's go deeper and add a hidden layer with three neurons, each containing a logistic activation function, to the previous network. This creates the network structure shown in Figure 12.2.

The idea is the neurons in the hidden layer will "learn" new and different *representations* of the input data that make the prediction task easier. Let's look at the top neuron of the hidden layer (that's the layer in the middle of the figure, comprising three neurons). After being trained on historical data, suppose the top neuron "learns" to appreciate that one's GPA, major, and number of extracurriculars are important factors to getting an offer. That means an applicant with a high GPA, many extracurriculars, and a difficult major would make this neuron "fire" a signal that represents a new feature within the data. We might call this feature *achievement*. From a mathematical sense, this means the weights for GPA, major, and extracurriculars are "large" within this neuron's activation function.

Likewise, the middle neuron might pick up on a combination of years in school and number of extracurriculars as firing a signal for *experience*, and the bottom neuron could fire if the student had the appropriate *skillset*. This is all hypothetical, of course—much like in PCA, we are categorizing features based on the combination of fields that appear to have the most effect on them.

To recap, the four original data inputs go into the hidden layer and come out as three new features. The features—achievement, experience, and skillset—then become the inputs into the final neuron, which takes a weighted combination of those inputs, runs it through yet another activation function, and produces a prediction.

From the computational view, the network can be thought of as a series of logistic regression models in each neuron.[7] Within the hidden layer, there

---

[7]We are making concessions here. If the activation is not the logistic function, then this statement is not true.

are three logistic regression models, each weighing the contributions of GPA, year, major, and extracurriculars differently. (For visualization purposes, we did not fully connect all inputs to the hidden layer; we're ignoring the connections that would have small, inconsequential weights.) The outputs of these three models then become the inputs into the final neuron, where a weighted combination of those inputs produces the final output.

This creates equations within equations—a nesting Russian matryoshka doll in math form. Let's give you an idea of what this would look like.

The "outer" function is the activation in the last layer of the network. For the network in Figure 12.2, it would be:

$$\text{Probability of Offer} = \frac{1}{e^{-\left(w_1 * Achievement + w_2 * Experience + w_3 * Skillset + b\right)}}$$

But each feature in this equation—achievement, experience, and skillset—are *separate equations*. If we replaced just *Achievement* in the preceding equation, we'd get (brace yourself)

$$\text{Probability of Offer} = \frac{1}{e^{-\left(w_1 * \left(\frac{1}{1+e^{-\left(w_{11} * GPA + w_{21} * Year + w_{31} * Major + w_{41} * EC + b1\right)}}\right) + w_2 * Experience + w_3 * Skillset + b\right)}}$$

And that's if we just replaced *Achievement*! We haven't replaced the others, but this hopefully drives home the point we made earlier: neural networks are giant math equations.

The consequence of this inception-like structure is a huge equation, with many parameters, that takes the input dataset and combines it in numerous ways. It's the layering of these functions that allows the network to identify more complex representations in the data, which creates the potential for more nuanced prediction.

And just as thoughts are complicated to describe, so too are the neural networks. Which is to say, in practice, the hidden layer likely won't produce interpretable representations for humans as we've shown here (achievement, experience, and skillset). Worse, it will get even more complicated as you add more layers and neurons. Sometimes these models are described as black boxes because of how little they make sense when they are layers and neurons deep.

So when explaining neural networks to others, you don't have to get caught up in dramatic comparisons to the human brain. More realistically, neural networks are large math equations, typically used for supervised learning tasks (classification or regression), that can find new representations of the input data to make predictions easier.

What then is deep learning?

APPLICATIONS OF DEEP LEARNING

Deep learning is the family of algorithms using the artificial neural network structure with two or more hidden layers. (In other words, it's an artificial neural network with better branding.) The idea of going deep (or as we visualize it in Figure 12.3, wide) with a neural network is to stack hidden layer upon hidden layer, where the outputs of one layer become the inputs to the next. At each layer, new abstractions and representations of the data are realized, effectively creating increasingly subtle features from the input dataset.

It's a complicated process that hasn't always been easy to do. In 1989, researchers led by Yann LeCun[8] created a deep learning model that took handwritten digits as inputs and automatically assigned the appropriate numerical label as an output. The goal was to automatically recognize zip codes on mail.

The network had over 1,200 neurons and nearly 10,000 parameters. (Think about that for a second. The model in Equation 12.2 has only 5 parameters.) His team needed access to thousands of handwritten digits with labels to learn from. All of that running on 1980s technology.

State-of-the-art computational power, a large labeled dataset, and patience would prove to be the formula for deep learning's success. But while research progressed in deep learning, it's fair to say groundbreaking results stalled for several years because (1) training a deep neural network was painfully slow even on the fastest, most expensive computers at the time; and (2) access to stockpiles of labeled input-output data was limited. Patience could only go so far.

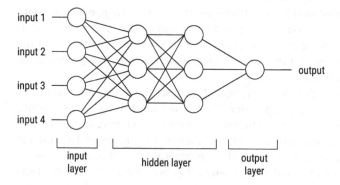

FIGURE 12.3   A deep neural network with two hidden layers

[8]LeCun, Y., *et al.* (1989). Backpropagation applied to handwritten zip code recognition. *Neural computation, 1*(4), 541-551.

But in the 2010s, the confluence of big datasets (thanks to the Internet), algorithmic improvements (like better activation functions than the logistic function), and computer hardware known as graphical-processing units (GPUs) started a revolution in deep learning. GPUs gave a hundred-fold improvement in training time.[9] Suddenly, the process of learning the thousands of parameters within a deep neural network went from weeks or months to a matter of hours or days. Deep learning victories have snowballed ever since, especially with unstructured data like text, images, and audio—from identifying and labeling faces to transcribing audio into text.

The Benefits of Deep Learning

Before we discuss how deep learning can handle unstructured data, let's talk more about why deep learning is different from the algorithms you're familiar with. We've touched on a few reasons already: the hidden neurons can generate new and subtle representations of a dataset, model interactions, and non-linear relationships, thereby finding nuance other methods might miss.

From a practical standpoint, this can be incredibly helpful to data workers because it reduces the time for manual *feature engineering.*

Feature engineering is the process of combining or transforming the raw data into new features (new columns) in a dataset using subject matter expertise. For example, in a dataset that predicts if someone will default on a loan, dividing the two inputs of household income and housing price to create an affordability metric—housing price/household income—may improve the performance of a model. But the process can be time-consuming and vague. Deep learning, within its hidden layers, can often automate this feature engineering process by creating representations of the data that are more amenable to the predictive task.

And with more data and larger, deeper networks, the automated feature engineering and layering upon layering of neurons might reveal increasingly complex and rich representations within the data that improve its performance as it learns from larger datasets. This is shown in Figure 12.4.

---

[9]See the article "From not working to neural networking" at https://www.economist.com/news/special-report/21700756-artificial-intelligence-boom-based-old-idea-modern-twist-not"

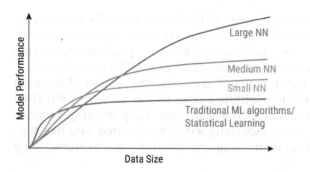

FIGURE 12.4 Theoretical performance curves of traditional regression and classification algorithms compared to small and large neural networks as the size of labeled data gets larger.[10]

The figure shows theoretical performance curves of different algorithms and how traditional methods (think logistic and linear regression) can stall in performance, even as the size of the labeled training data continues to grow. There's only so much signal the linear methods can capture. But the idea of going deeper and deeper with the neural network architecture is to continue to squeeze more information and predictive performance out of the data. And as the data increases in size, the performance of large, deep neural networks might continue to improve. From a practical matter, of course, there's a ceiling—every dataset has a limit. All the juice will eventually be squeezed from the lemon.

There's a huge caveat to Figure 12.4. The model performance will only increase if there is meaningful signal or information in the data. And this is no guarantee.

Deep learning, with its automated feature engineering and ability to pick up subtle patterns in data, performs well on problems of perception. In the next sections, we'll provide insight to how this works.

How Computers "See" Images

In the previous chapter, you learned how a computer "reads" text. In this section, you'll learn how computers "see" images and get a preview to how deep learning works in the field of computer vision.

---

[10]Image from: lilianweng.github.io/lil-log/2017/06/21/an-overview-of-deep-learning.html and inspired by an image in Ng, A. (2019). Machine learning yearning: Technical strategy for ai engineers in the era of deep learning. *Retrieved online at* mlyearning.org.

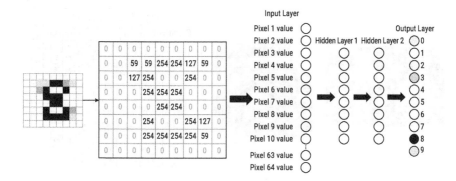

FIGURE 12.5 How a grayscale image "looks" to a computer, and how that data would feed into a deep neural network. Darker shades in the output layer indicate the most probable guess.

Figure 12.5 shows how a simple grayscale image—a handwritten digit—looks to a computer.[11] Each pixel of the image has been converted into a value ranging from 0 (representing white) and 255 (black) and every shade of gray in-between. Figure 12.5 is a low resolution 8-by-8 pixeled image, represented as a matrix with 64 values ranging from 0 to 255. Humans see a handwritten digit on the left—the computer sees a spreadsheet of numbers in the middle.

Now, imagine a database of several thousand examples: handwritten digits between 0 and 9 of various styles and penmanship. If asked to read and identify the digits, humans, including children, could accomplish this effortlessly. But how would a computer perform image classification?

We could apply a deep learning algorithm on this dataset that could learn from thousands of handwritten numbers. We can think of the neurons in the hidden layer "firing" if, say, a loop was present in the digit (a 0, 6, 8, or 9), or perhaps a vertical line (1, 4), or a horizonal line, (2, 4, 7) or even some mixture of the three.

We cheated again, here, to give a sense of what the neurons *might* represent. But as established already, hidden layers are often hard to interpret, and they might produce representations that don't make immediate sense. But the conceptual idea holds. The inner neurons can indeed pick up on patterns in the digits, but they only make sense mathematically—not visually.

---

[11]Automatically uncovering handwritten digits is a rite of passage when studying deep learning. This is the problem Yann LeCun solved in 1989. Today, the process can be done on a laptop. A database of handwritten digits is available here: yann.lecun.com/exdb/mnist.

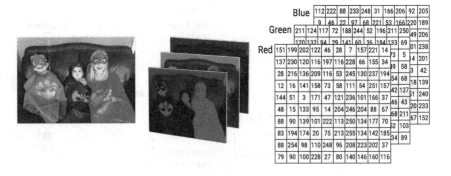

FIGURE 12.6 Color images are represented as 3D matrices for the pixel values corresponding to red, green, and blue values.

Convolutional Neural Networks

Let's now move to a more advanced way to analyze images, convolutional neural networks, which are used by researchers to build image classification systems on larger images (more pixels) and those with color.

We'll start with how a computer "sees" a colorful image. Each pixel in a colored digital image is made up of three colors—red, green, and blue. We call these color channels. The red channel contains a matrix of values, ranging from 0 (no red) to 255 (red), and likewise for the green and blue channels. So, instead of one matrix of numbers, you have three, as shown in Figure 12.6.

Consequently, a 10-magapixel image would contain 30 million numbers (a red, blue, and green number for each of the 10 million pixels). And if these 30 million inputs were fed into a neural network with a hidden layer that had 1,000 neurons, your computer would need to learn a whopping 30 billion weight parameters.[12] If you don't have access to the world's fastest supercomputer (heck, even if you did), you'd better find a better way to handle this.

How do researchers and deep learning practitioners do it? By using a process called convolution. Convolution is the mathematical equivalent of analyzing a picture with a series of magnifying glasses, each with a different purpose. As you move left-to-right, top-to-bottom over an image with the magnifying glasses, you'll notice many localized patterns: lines, corners, rounded edges, textures, and so on (see Figure 12.7). Convolution does this mathematically—performing calculations on a localized set of pixel values,

---

[12]Each of the 1,000 neurons in the hidden layer would be a weighted sum of the 30 million input values.

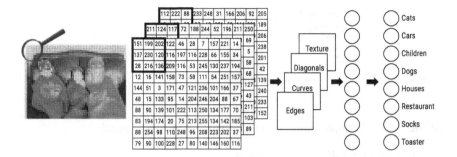

FIGURE 12.7 Convolution is like a series of magnifying glasses, detecting different shapes in an image, which feed into the hidden layers of a neural network for classification.

finding edges (0 values next to large values, for example), and other patterns. Then, to reduce the sheer size of numbers involved in the process, it "pools" the numbers together to find the most distinctive features.

After convolution finds local patterns like horizontal or diagonal edges, the neurons in the hidden layer start to piece the important edges back together (in a mathematical sense) and filter out information irrelevant to the target output. It twists and turns the data in such a way that the network can learn to tell if children are in a photo, or the difference between your authors' faces in a different photograph. Or in the case of self-driving cars, the difference between parked cars from moving cars; pedestrians from construction workers; and stop signs from yield signs.

The process of convolution not only reduces the number of values that go into the familiar neural network structure (remember, you don't want to estimate billions of numbers if you can avoid it), it also "searches" for similar features across images. Unlike structured datasets where the location of a feature is known and fixed in a column, features in images not only have to be analyzed, but located. This is why social media can find your face in a picture no matter where you're hiding.

Deep Learning on Language and Sequences

Deep learning has also powered advancements in language and sequences using a structure called a *recurrent neural network*. As you recall from the previous chapter, traditional text analytic methods fail because they ignore word order. Everything is just thrown into a bag of words.

But word order clearly matters. Consider the following two sentences with the word "orange." Can you predict the final word in each?

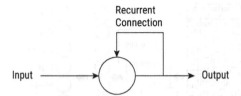

FIGURE 12.8 A simple representation of a recurrent neural network

1. At breakfast, I like to drink orange _____.
2. My cousin in California lives in Orange _____.

Your brain probably filled in the missing values without missing a beat: juice and County. The moment you reached the missing word in the first sentence, the words "breakfast" and "drink" were in your short-term memory. Clearly the answer was juice. Likewise, your memory of "California" and "lives in" revealed the answer to the second sentence as California's Orange County. In both cases, your brain held on to past information while processing new. A recurrent neural network[13] is the computational equivalent.

Figure 12.8 shows a simple recurrent neural network where the output loops back into the network, creating a "memory."

For this "what's the next word?" text classification problem, thousands or millions of input-output pairs of intersecting sequences of words form a training set. For instance, the input "At breakfast I" would be mapped to the output "breakfast I like." As the system scrolls through the sequences of words in a sentence, it also "remembers" the earlier words. Thus, when the network sees the input "to drink orange," the output will likely be "drink orange *juice*," if the historical data contained sentences with the phrase "orange juice" or "drink juice."

You can see how deep learning algorithms like this can be used to help you more quickly craft responses and correct your grammar as you type. This technology has been available in Google's "Smart Compose" in Gmail since 2018; it suggests text to finish your _____, and it's powered by recurrent neural networks.[14]

Now that we've discussed the technical details of deep learning, let's talk about it in practice.

---

[13]There are several types of recurrent neural networks. The most popular is called the Long short-term memory (LSTM) network.
[14]www.blog.google/products/gmail/subject-write-emails-faster-smart-compose-gmail

DEEP LEARNING IN PRACTICE

Admittedly, it's hard not to be excited about deep learning. We're only scraping the surface of its potential. And today's biggest thinkers are making a compelling case that it will drive much of our future. But alas, this excitement can blind us from the challenges that remain—indeed, the challenges that have always remained—when working with data.

Do You Have Data?

As amazing as deep learning is and sounds, perhaps the biggest hurdle to companies is this: you might not have enough labeled training data. Like the quote says at the beginning of the chapter, data "is the raw material that powers our intelligent machines, without which nothing would be possible" —yet we see it time and time again (and we've mentioned it several times already) that many businesses want to rush to try deep learning without having enough labeled data for their specific application.

Deep learning and AI expert Andrew Ng, articulated the data challenge in this way:[15]

> A major challenge to taking advantage of AI throughout the economy is the sheer amount of customization needed. To use computer vision to inspect manufactured goods, we need to train a different model for each product we want to inspect: each smartphone model, each semiconductor chip, each home appliance, and so on.

And each of those models would require their own, likely large, set of labeled images.

Transfer Learning (or How to Work with Small Datasets)

If you have *some* labeled data, perhaps hundreds of images but not thousands, your team might find luck with something called *transfer learning*.

The idea of transfer learning is to download a model that's been trained to identify everyday items (balloons, cats, dogs, etc.).[16] That is to

---

[15]blog.deeplearning.ai/blog/the-batch-ai-researchers-under-fire-rl-agents-in-danger-bias-in-synthetic-data-one-neuron-to-rule-them-all

[16]Many practitioners will use models trained on the ImageNet database (en.wikipedia.org/wiki/ImageNet) for transfer learning.

say, the thousands of parameter values in the network have been optimized to work with a group of images. Recall the early, shallow layers of neural networks trained on images learn generic representations like shapes and lines. The later, deeper layers piece those edges and lines together to form the expected output image.

The idea behind transfer learning is to pop off the final few layers— the layers that learn how lines and edges form cats and dogs, for example—and replace it with new layers that, with a fresh round of training, learn how those shapes combine to form the outlines of tumors in medical images. Note, transfer learning may reduce the number of labeled images by a factor of 10, but it won't bring it down to dozens.

Is Your Data Structured?

The mystique around deep learning is due in large part to its predictive performance on perceptual data: images, videos, text, and audio. Data that we can comprehend, get excited about, and use the results without having to look at a spreadsheet. With structured data, your typical rows and columns, deep learning might still help you improve performance, but often it doesn't.

If your data workers rush to deep learning as a life raft when trying to build a supervised learning model on a structured dataset—"All else has failed! Let's try deep learning!"—we predict disappointment.

With structured data, deep neural networks often lose to tree-based methods (from Chapter 10). There are exceptions, no doubt, but if your model accuracy is embarrassingly low with tree-based models, your time will be better spent focusing on cleaning up issues in your data and figuring out if the problem you're trying to answer is even reasonable. (Remember, just because you have labeled data does not mean you will find a connection between your inputs and outputs.) Deep learning works well exploiting relationships between inputs and outputs *if* such a relationship exists. It cannot, however, generate something from nothing.

The quality and richness of your data still matters.

What Will the Network Look Like?

We made setting up deep neural networks sound easy, but there are dozens of decisions to be made. For instance:

- How many layers should the network have?
- How many neurons in each layer?
- Which activation functions should you use?

We won't answer those questions here (there are plenty of excellent books out there that do—see the sidebar), but we wanted to make it known that data workers will budget significant time, possibly weeks, to experiment with these parameters and the general architecture of the network. And, when building a large network, be careful not to overfit your data (all the lessons from Chapters 9 and 10 still apply!).

Deep Learning for Practitioners

If you want to learn how to build deep learning models yourself, we highly recommend the series by François Chollet, whose books teach the Keras deep learning library for the R and Python languages.

- Chollet, F. (2018). *Deep learning with Python*. New York: Manning.
- Chollet, F., & Allaire, J. J. (2018). *Deep Learning with R*. New York: Manning.

ARTIFICIAL INTELLIGENCE AND YOU

As we wrap up this chapter, we need to briefly talk about artificial intelligence (AI) and its broader implications. For the purposes of being a Data Head, you must understand there are two types of artificial intelligence. The first is *artificial general intelligence* (AGI), the idea of complete human cognition. Insert your favorite science fiction movie reference here. But rest easy, little progress is being made in AGI—at least not enough that you should worry.

Significant progress, however, is being made in *artificial narrow intelligence* (ANI): computer systems that do one thing well, like facial detection or speech translation or fraud detection. That is to say, ANI works because machine learning works. Indeed, you are safe to assume *AI is machine learning*. When you and your fellow workers talk about AI, and when vendors come to pitch you on AI—we're all really talking about machine learning. And if the problem involves perceptual, unstructured data, they are talking about deep learning. Machine learning is a subset of AI, and deep learning is a subset of machine learning (see Figure 12.9).

Some people use AI more liberally than others. For instance, a movie recommendation system is often referred to as AI by society at large, when it would be more accurately described as machine learning or statistical learning. Why does that matter? Because realizing "AI," as you hear it in the news, requires large, curated datasets from humans like you and me opens up honest discussions

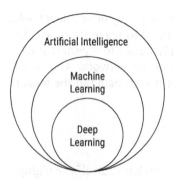

FIGURE 12.9   Deep learning is a subfield of machine learning, which is a subfield of artificial intelligence.

about data quality, variability, possible target leakage, overfitting, and a host of other practical concerns. AI is reinforcing patterns from data collected in the past. It's not about creating something resembling human consciousness.

Big Tech Has the Upper Hand

However, this dichotomy exists because Big Tech again has the upper hand here. For years, they've been quietly collecting labeled training data to train machine learning and deep learning models.

Remember when you clicked your Facebook photos years ago? So did millions of other people, giving Facebook a pile of images (inputs) with the locations of faces (outputs). Deep learning can now draw a box around your face and identify if it's you or a friend. And those annoying prove-you-are-human prompts when you log on to certain sites ("Select all boxes that contain a street crossing") are being used to train the deep learning networks behind self-driving vehicles.[17] Perhaps you might rethink hopping into a self-driving car until websites stop asking you to identify stop signs.

Data collection is the lesser known part of deep learning. It's certainly not as glamorous as talking about the human brain and automatic image classification. But, if you are wondering how your company can begin to find value with deep learning—or more broadly with machine learning—getting labeled data would be the first step. If you have data, for example images, that need to be labeled and you don't want to do it, fear not. An industry has been built up

---

[17]hackernoon.com/you-are-building-a-self-driving-ai-without-even-knowing-about-it-62fadbfa5fdf

around this idea and you can pay pennies on the dollar to have people label your data for you. And a future in which you could receive affordable access to the datasets you require probably isn't that far off.

Ethics in Deep Learning

Your authors are not ethicists, so we aren't necessarily the right people to lead this discussion. But, either way, a Data Head doesn't have to lead this discussion to be part of it. As you are on the front lines of working with data, you must be a faithful participant in how it's used.

Data is growing faster than our ability to articulate the problems it creates. On one level, this makes sense in a rhetorical and common critique of all new technologies. But using data raises even larger problems. Both because we erroneously believe it always represents an unshakeable ground truth and because the algorithms fascinate us by how much they appear to replicate our own decision-making abilities.

And while we've stressed that the algorithms themselves aren't copies of human thinking, the reality is that their applications can appear human enough to fool us. Hackers, for instance, have used a type of deep learning called Generative Adversarial Networks (GANs) to create "deep fakes" of people's faces. This would allow them to superimpose a fake face on a real person, to make it look like they did something they did not. News about fake events could be tweeted with realistic headlines generated from actual headlines. These are the ways in which this technology can fool us.

And, on an even deeper level, we should be careful about what human evaluations we attempt to replace with deep learning. For instance, how useful is a tool for a judge to predict the recidivism rate of a previous offender using AI?

As we've described, the largest critique of deep learning is the sheer confusion of what is going on behind the scenes. It's hard to explain a giant math equation with millions of parameters, yet these equations might be used when sentencing criminals; or as a security feature on a phone (Apple's iPhone Face Scan); or to slam on your car's brakes or swerve to miss that deer from earlier in the chapter.

Moreover, often what we are modeling aren't just data points—they're people. Aspects of their identity are coded and labeled. When we receive the data, it might not have much meaning to us. But if we accept that data grows faster than our ability to articulate the problems it creates, we must not assume society has already approved our use of data. That we can collect certain features and run algorithms doesn't always mean we should. And while we've given you the tools to understand well-constructed deep learning applications, you cannot assume every application has done them correctly. Even within your own organization, do not be immediately impressed by claims of deep learning solving the problems. Ask not just to see the data and algo-rithms—but *ask who does this result affect? Am I OK with that?*

All of which is to say, that while machines are getting smarter, you must do so as well. Do not take for granted your own role in using data to make outcomes better for business and society.

CHAPTER SUMMARY

In this chapter, we brought together many of the lessons of previous chapters to explain how deep learning works. Remember, deep learning is built on top of artificial neural networks, artificial neural networks are made up of neu-rons, and each neuron comprises an equation called the activation function. Each layer feeds into one or more neurons. These layers become sub-functions that flow into the final layer, that, when all put together, becomes one large (and impressive!) mathematical equation, which serves as a predictive model.

Deep learning represents an exciting new chapter for machine learn-ing. The ability to run more complicated models has become easier and less expensive by the day. Yet, despite this promise, we must set our expectations of deep learning appropriately to start. Deep learning is good for perceptional problems like classifying images and text made up of high-quality, correctly labeled data. It's not always the best choice to solve smaller problems on struc-tured data.

In the end, humans run models. Do not let the mystique surrounding deep learning lead you to think that it's smarter than you—or that your use of it is neutral. At the end of the day, it's your work, and you must feel comfort-able to stand behind it.

Ensuring Success

In Part IV, you'll discover how to make the most out of your Data Head journey by learning from others' mistakes, both technical and human.
Here's what we'll cover:

Chapter 13: *Watch Out for Pitfalls*

Chapter 14: *Know the People and Personalities*

Chapter 15: *What's Next?*

Watch Out for Pitfalls

"The first principle is that you must not fool yourself, and you are the easiest person to fool."

Richard P. Feynman, Nobel Prize-winning Physicist

Thinking, speaking, and understanding data is—in many ways—about knowing what mistakes can happen if you don't have your wits about you when working with and interpreting data. Some pitfalls will be easy to fix, but they're hard to find if you don't know where to look. And if you're not careful, they can turn into major data disasters like those introduced throughout this book (think *Challenger* and the 2008 housing market collapse).

Our goal in this chapter is to remind you of the pitfalls you've learned about and introduce you to a handful of other common pitfalls that, if you're not careful, can derail your work or (worse) convince you of something that isn't so.

Before we begin, let's take a moment to acknowledge that it's easy to complain about others' mistakes. And we admit, data pitfalls and blunders are fun to discuss. While we encourage you to cast a skeptical eye on the work completed in your industry, let's also realize that change comes through empathy and encouragement. Honest mistakes happen, and to be sure, your authors came by the knowledge in this chapter the hard way. Thus, let's acknowledge that most of the pitfalls have no nefarious intent or bad faith behind them. Many are likely caused by people simply not knowing what can go wrong. This chapter brings those issues to the surface.

BIASES AND WEIRD PHENOMENA IN DATA

Bias is a complicated subject that cascades across various disciplines. We understand bias as the lopsided (and sometimes even inconsistent) favorability given to ideas and concepts by individuals and reinforced in groups. In this section, we'll discuss common biases seen in the world of data, as well as phenomena where the first glance at data can convince you of one thing, but a second look reveals another.

Survivorship Bias

Suppose an investment company launches dozens of mutual funds the same year, each containing a random assortment of stocks. If a fund fails to beat its performance benchmark in a certain time frame (for example, if the S&P 500 returns 10% and one of the funds only returns 3%), it's silently discontinued. After several years pass, only the top performing mutual funds—the survivors—remain, with an impressive return to boot. At which point you, a potential investor, comes along with money to invest. You're shown data advertising the year-over-year market-beating performance of the company's funds.

Would you feel comfortable investing?

Perhaps. Businesses prune poor performers in a variety of activities, which isn't inherently bad. However, pretending the poor performers never existed *is bad*, as this creates bias. In this example, you weren't presented data on the "bad" or unlucky funds because they were closed. This will bias the company's performance upwards, making you believe the company has expert stock pickers when one plausible explanation is simply luck.

This is an example of *survivorship bias*, the "logical error of concentrating on people or things that made it past some selection process and overlooking those that did not, typically because of their lack of visibility."[1]

A classic example of survivorship bias is that of statistician Abraham Wald, who attempted to minimize losses to the Allied bomber fleet during World War II. Planes that survived nasty dog fights would return with serious damage and bullet holes to their wings. The original idea was to reinforce planes where this damage was pervasive. However, it occurred to Wald that this reflected a survivorship bias. In fact, they were learning from the planes that *returned*. But what about those that did not? What does the damage pattern say about them?

---

[1]Survivorship Bias: en.wikipedia.org/wiki/Survivorship_bias

Wald's seemingly counterintuitive recommendation: armor the sections where the returned planes had the *least* damage. Why? Because planes hit in those areas never made it home.

Regression to the Mean

Regression to the mean is a phenomenon that sounds simple enough: extreme values of random events are often followed by not-so-extreme values. This observation was originally coined the "regression towards mediocracy" in 1886 by Sir Francis Galton[2] who noticed tall parents' children were not as tall as them (the children *regressed*) and short parents' children were not as short. What he was really noticing was that there is a natural, underlying stability in the height of humans and their offspring. In either case, extreme values (short and tall) were usually followed by not-so-extreme values (not as short or tall).

While that example might seem obvious, regression to the mean has broader implications in inference. If you do not take a holistic, bird's-eye view of all available data, certain observations can feel extreme. The bias then is to act on these extreme events, not considering that a more predictable event—closer to the true underlying mean—is on the horizon, whether you act on the data or not.

Consider a National Football League (NFL) player who has an outlier year and is featured on the cover of the popular video game, *Madden NFL*, only to regress to a less-than-stellar performance the following year. This has been deemed the "Madden Curse."[3] But we know it as regression to the mean. Or imagine a traditionally decent employee who has a down year and receives poor performance reviews. They're put on a remediation plan, and the next year, the employee's performance rebounds. The manager attributes the employee's improved production to the manager's sage wisdom and guidance, but performance probably would have improved anyway, due to regression to the mean.

The message of regression to the mean is this: Don't put faith in outliers. Luck—good or bad—won't last forever.

Simpson's Paradox

Another phenomenon to look out for is known as *Simpson's Paradox*, a potentially disastrous pitfall with observational data (which is most of the

---

[2]Galton, F. (1886). Regression towards mediocrity in hereditary stature. *The Journal of the Anthropological Institute of Great Britain and Ireland, 15*, 246–263.
[3]en.wikipedia.org/wiki/Madden_NFL#Madden_Curse

data you'll work with). Simpson's Paradox happens when a trend or association between variables is reversed after the addition of a third variable. With Simpson's Paradox, you must avoid two data mistakes: thinking correlation is causation *and* having the wrong correlation.

Consider the data in Table 13.1 from a 1986 study on two different types of surgical techniques to remove kidney stones.[4] A review of medical records showed that a new, minimally evasive procedure for kidney stone removal had a higher success rate (83%) than the traditional procedure (78%). The results were statistically significant and by all accounts seemed conclusive.

Unfortunately, Simpson's Paradox was lurking in this data. A further review of the data showed the result is reversed when broken out into kidney stone size. As it turns out, the traditional procedure had higher success rates on both patients with small kidney stones (< 2 cm diameter) and on patients with large stones (≥ 2 cm diameter). The breakdown is shown in Table 13.2.

How can that be? Because the newer procedure was tested on many patients with small kidney stones (presumably the easier cases), while the traditional procedure was primarily used on the patients with larger kidney stones. Even though the traditional procedure performed better on small kidney stones (at 93%), the new procedure was performed on a lot more patients with an 87% success rate. The new procedure's overall success rate, then, is weighted more toward 87%. In Table 13.2, we can see the overall success rate for the traditional procedure (78%) was weighted more toward the success rate of patients with large kidney stones (73%). The new procedure performed worse on this group, but it was on too few patients to move the needle of its overall success rate. Confused? It's okay—that's why it's called a paradox.

What is the best way to mitigate the risk for Simpson's Paradox? Randomly split observations into each treatment group to avoid any confounding. In other words, collect experimental data.

TABLE 13.1 Success Rates of Surgical Techniques to Remove Kidney Stones

Treatment	Overall Success Rate
Traditional Procedure	78%
New Procedure	83%

---

[4]The example first appeared in: Julious, S. A., & Mullee, M. A. (1994). Confounding and Simpson's paradox. *Bmj, 309*(6967), 1480–1481. We learned about it in the excellent book: Reinhart, A. (2015). *Statistics done wrong: The woefully complete guide.* No Starch Press.

TABLE 13.2 Simpson's Paradox Lurking in the Success Rates of Surgical Techniques to Remove Kidney Stones

Treatment	Small Kidney Stones	Large Kidney Stones	Overall
Traditional Procedure	93%	73%	78%
New Procedure	87%	69%	83%

Confirmation Bias

Confirmation bias is a potential pitfall in any data project. It happens when data and results are interpreted in a way that confirms one's already held beliefs, while conflicting evidence that doesn't align with prior beliefs is cast aside as moot.

While it's easy to accuse senior executives, politicians, and business stakeholders of confirmation bias, it's much harder to look inward at ourselves. But for many data teams, confirmation bias is almost a way of life. That's because the team is stood up to find evidence that executives are making the right moves—moves that may have already been made before much data was analyzed. For these teams, at least some of the data work exists to feed the confirmation bias machine. This is not an easy or fulfilling place to be in, but Data Heads should strive to rise above any confirmation bias and report findings truthfully. If not, the team could succumb to a confirmation bias apparatus used to justify decisions rather than as a proper data team designed to understand all decisions available, without undue business or political pressure.

Effort Bias (aka the "Sunk Cost Fallacy")

Effort bias refers to the desire to keep going when vast amounts of time, money, resources, and effort have been sunk into a project. Once this happens, it's hard to turn your back on the results even if you recognize early on that:

- You don't have the right data for the project.
- You don't have the right technology for the project.
- The original project scope fails to capture the underlying merits of the project.

Some companies would rather you deliver something, *anything*—given the time, effort, and attention already paid. This type of pressure is the fertile ground upon which many of the biases listed previously are formed.

Algorithmic Bias

As more decisions are automated with machine learning, we are becoming aware of a type of embedded prejudice already baked into the data and computational world. This is called *algorithmic bias*.[5] Although researchers and organizations have only recently begun to take a much closer look at its origins and impacts, such bias has always existed in the data. It's often the product of the status quo, and it can be hard to detect en masse until the status quo is fundamentally challenged. However, by bringing an awareness to your work, you can detect it much sooner.

Think back to the example in the previous chapters where we shared data about intern applicants and tried to predict if they received an interview offer. If the dataset included gender, coded as a categorical variable, and more men than women were historically offered interviews, every algorithm would detect and exploit this relationship and give a greater weight of predictions to men. To an algorithm, it's all 1s and 0s, but Data Heads should know this bias happens even at top tech companies at the forefront of machine learning like Amazon.[6]

Be advised: it's not just that algorithmic bias can happen anywhere, however good (or neutral) your intentions. It is already happening. No model's predictions represent the final truth. All results are products of assumptions. And you must operate as if all observational data has bias baked in—because it does. When models make predictions, they perpetuate and reinforce underlying bias and stereotypes already manifest in the data. You cannot rely on a change in mindset to start digging into the biases in your own work. This work should begin today.[7]

Uncategorized Bias

What we covered in this section is not a complete list of biases, paradoxes, and weird data phenomena. In fact, we want you to be on the lookout for pitfalls that have no category. If you looked only for specific types of bias or

---

[5] Algorithmic Bias: en.wikipedia.org/wiki/Algorithmic_bias

[6] A 2018 article in Reuters, titled "Amazon scraps secret AI recruiting tool that showed bias against women," details how their learning algorithms penalized resumes with the word "women's" and the names of all-women colleges. www.reuters.com/ article/us-amazon-com-jobs-automation-insight/amazon-scraps-secret-ai-recruiting-tool-that-showed-bias-against-women-idUSKCN1MK08G

[7] The Brooking Institution has a resource to get you started: www.brookings.edu/ research/algorithmic-bias-detection-and-mitigation-best-practices-and-policies-to-reduce-consumer-harms

logical fallacies, you might miss other biases that aren't as prominent if only because we have not defined them as a society. But it doesn't mean such pitfalls aren't there.

THE BIG LIST OF PITFALLS

Now that you're familiar with some general biases and cognitive traps when working with data, let's talk about more specific pitfalls to avoid in data projects. We've organized this big list into two categories: (1) statistical and machine learning pitfalls and (2) project pitfalls.

Statistical and Machine Learning Pitfalls

This section contains a list of statistical and machine learning traps, many of which we've discussed previously.

> **Thinking correlation is causation**: Resist the temptation to build a causal narrative around correlated variables. A company's increase in sales can be correlated with more YouTube ad views, but the increased ad time may not have *caused* the increase in sales. A good rule-of-thumb is to avoid talk of causality unless you've specifically designed your data collection and analysis around finding a causal relationship (i.e., you're using experimental data!). Chapters 4 and 5 discussed these ideas.
>
> *p*-**hacking**: Suppose an article proclaims, "People who drink too much coffee have an increased risk of stomach cancer. The result is statistically significant at the 0.05 significance level."[8] Recall from Chapter 7 that signals in data at the 0.05 level will be false positives 1 in 20 times. *p*-hacking is the process of testing multiple patterns in data until you spot a statistically significant *p*-value. A link between coffee and stomach cancer would be less worrying if you later discovered researchers also explored correlations between coffee consumption and brain cancer, bladder cancer, breast cancer, lung cancer, or any of the 100 types of cancer. By sheer coincidence, five of these would show a statistically significant *p*-value, even if no relationship existed. Note, *p*-hacking is a type of survivorship bias, as only the significant *p*-values are reported.
>
> **A non-representative sample**: Election polls that do not represent the voting population will be wrong. A survey of your company's social

---

[8]This is only an example. Your authors do not research cancer.

media visitors may not reflect what most of your customers think. Feel empowered to argue with the data (Chapter 4) because setting policy or basing decisions on a sample of data that does not represent the population it will influence can lead to serious errors. Worse, the data can provide a false comfort, tricking you into thinking you're making a data-driven decision when perhaps no data would have been better than the bad data you're working with.

Data leakage: Do not train a model on data that will not be available at prediction time. You will be fooled into thinking your team has an excellent model when it may be utterly useless. Predicting if a visitor on your website will buy a product is easy if you know they applied a coupon code at purchase. Data Heads must check that each feature in a model is present when a decision is needed (Chapters 9 and 10).

Overfitting: Recall that models are simplified versions of reality. They use what we know to help us predict what we don't. But when the model seems to perform well on data it's seen before but can't predict new observations, the model can be said to be "overfit." In a sense, the model is "memorizing" the scenarios defined by the training data rather than "learning" from the training data to make predictions about the unknown (Chapters 9 and 10). Data Heads can prevent overfitting by splitting the data into training and test sets. Learn from the training set and see how well the model predicts on the test set.

Non-representative training data: This pitfall is using a "non-representative sample" to create a machine learning model. Models only know the data they were trained on. A model that learns from real estate data in Ohio to predict the sales price of homes in Ohio cannot predict the rental price of an apartment in New York City. Likewise, a voice-assistant smart speaker that is trained on audio samples recorded in a sound room may have trouble parsing commands in a loud home. To avoid this pitfall, Data Heads must think carefully about the circumstances in which their model will be used and collect training data to reflect those applications.

Project Pitfalls

In this section, we'll describe many of the project pitfalls that can happen to a data science project if you're not careful.

Not asking a sharp question (or solving the wrong problem): Even the subtlest ambiguities can cause misalignment and confusion between the data team, business team, and project stakeholders. Make sure everyone is crystal clear on the business problem being solved (Chapter 1).

Not adapting the question once it's already failed: Just as important as a crystal-clear business problem is acknowledging quickly when it can't be answered. Many data teams will find quickly the original question was lacking but will push forward because of external pressure. The question must be updated, or misalignments will occur.

Data is owned, not governed (i.e., data is hard to acquire): In some organizations, specific teams (like IT, Finance, or Accounting) own the data that you need to work with. While many of these organizations practice governance on paper, it can be used to keep your access away. Your company must understand that you can only do so much if data is restricted.

Data does not contain the information needed: Data might be easy to acquire and "tidy," but it might not contain the information you need to solve your problem. If data does not contain the information you need, try to collect better data that does.

Not relying on inexpensive and open source technologies: Before taking on that huge project around some new technology implementation, take a moment to prototype it first. It may very well be that investing in a data science platform to run the future of your operations will be a game changer for the team. But before anyone spends a dime, wouldn't it make sense to build a minimum viable prototype (MVP) in Microsoft Excel or by using a free, open source technology like R or Python?

Timeline is overly optimistic: Data projects often fail in unexpected ways. The preceding problems are not discovered until weeks into a project, and aggressive timelines lead to shortcuts and bad analysis. Project timelines should account for the inevitable data setback.

Inflated expectations of value: Businesses have been groomed to expect a lot out of data science, statistics, and machine learning. Be transparent with the value your project can bring, but don't oversell it. This can create backlash that hurts current and future projects.

Expecting to predict the unpredictable: Some tasks, no matter how much historical data you collect, can't be predicted. Documenting every spin of every roulette wheel in Vegas won't help you predict the next spin.

Overkill: Like you, your authors love data projects. So many of us are ready to dig into the next project idea. But what's often lost is perhaps deeply obvious: Data science, statistics, machine learning, and AI can solve many important problems in the world, but they can't solve every problem. One pitfall that is often lost in the technical details of working with data, statistics, and algorithms is *overkill*. You can have a classification algorithm help you identify business rules. But, sometimes, we already know and operate on a set of rules. And, in these cases, it's easier to have the humans around you *write them down*. Essentially, if your team can write the business rules to automate a process, your work is done. This is lost in the hype of the field right now. Machine learning sounds compelling to management, but sometimes, it's just overkill.

CHAPTER SUMMARY

In this chapter we covered both common biases and pitfalls. As we stated previously, this list isn't exhaustive. And you should come from the standpoint that no list is. Recall that data is growing faster than our ability to articulate the problems and opportunities it creates. If you accept this, then no list can fully capture what pitfalls people haven't yet fallen into. But this chapter gave you a place to start.

Projects fail all the time. And chances are, you will have at least one (but, most likely, more) project fail that you are associated with. Be open and transparent when failures happen, and pivot to new ideas if possible. Indeed, your experience will be your best teacher.

CHAPTER 14

Know the People and Personalities

"People worry that computers will get too smart and take over the world, but the real problem is that they're too stupid and they've already taken over the world."[1]

—Pedro Domingos, AI Researcher

In the previous chapter, you learned about common project pitfalls. In this chapter, we'll talk about the people and perceptions of roles when working with data, and how many projects fail not because of technology or data, but because of conflicting personalities and poor communication.

It's a lack of communication that drives many of the project failures described in this book. Our goal is to help you navigate communication red flags by understanding the different personalities involved. Thus, this chapter discusses the perceptions held by key personalities and shares scenarios about what happens when communication breaks down between data workers and business professionals. Understanding the roles of others and showing empathy will go far in bridging any communication gaps. This is how you become a Data Head.

In the next section, we'll explore more observations about data and business professionals and highlight scenarios when the communication gap kills projects. Then, we'll talk about different attitudes people have toward data—enthusiasm, cynicism, or skepticism.

---

[1]Quoted here: www.washington.edu/news/2015/09/17/a-q-a-with-pedro-domingos-author-of-the-master-algorithm

SEVEN SCENES OF COMMUNICATION BREAKDOWNS

When communication breaks down, you might see one of these seven scenes play out in a data project.[2] We've summarized them in Table 14.1. In the sections that follow, we go into more detail, presenting scenarios for each that might sound a bit too familiar.

TABLE 14.1 Seven Scenes of Communication Breakdown

Scene	Summary
The Postmortem	A senior data scientist is brought in to get a project back on track long after early warning signs. It's too little too late for the project.
Storytime	A smart analyst strips his presentation of technical nuance to satisfy the myth that they must explain stuff to higher-ups like they're children—and the analyst feels like they're betraying their role as a critical data thinker.
The Telephone Game	A preliminary statistic, manifest of code and data science work, is taken out of context and then shared so widely it loses whatever little meaning it originally had.
Into the Weeds	Results are so technical as to be rendered meaningless. The resulting deliverable is more self-indulgence than a true presentation on what happened.
The Reality Check	The data worker pursues perfecting an impractical solution and does not consider alternatives until challenged by authority.
The Takeover	A data scientist attempts to try to solve major underlying business problems without establishing team trust, rapport, or focusing on quick wins.
The Blowhard	A data scientist finds fault with virtually all work that isn't his. As a result, he is no longer sought to support projects.

The Postmortem

A high-profile project at a telecommunications company has stalled after six months of work.

The project team, consisting of one data scientist, was tasked with predicting customer *attrition*. The activity they were trying to predict was

---

[2]This section was inspired by the article "Data Science and the Art of Persuasion" by Scott Berinato (hbr.org/2019/01/data-science-and-the-art-of-persuasion), based on our own experience and the experiences of our business colleagues who were gracious enough to share their stories.

whether a customer would switch to a new cell-phone carrier in the next year. They developed a model that assigns scores to all its current customers using historic data: customer_1 has an 85% chance of switching carriers, customer_2 has a 10% chance of switching, etc.

On paper, the job is complete. Models can be run. Code has gone into production. But there's a small (well, maybe not so small) problem: the model is nowhere near as accurate as the team promised to the business stakeholders.

The project lead, having tap-danced over several ongoing concerns from the data scientist for the past few weeks, assumed issues were minor technical problems that could be quickly fixed. (Computers can do everything, right?) But the problems are much bigger than anticipated and the leader is getting nervous. So, they bring in an even more senior data scientist to take over the project.

But it's too late.

Hundreds of decisions have been made at this point, and the expert cannot even begin to untangle the mess, especially not in the final week before the team presents results to senior management. The expert not only reiterates the concerns of the team's data scientist but adds even more to the list.

While pulling another 12-hour day to salvage what she can of the project, the senior data scientist is reminded of the eminent statistician R.A. Fisher's quote, "To call in the statistician after the experiment is done may be no more than asking him to perform a post-mortem examination: he may be able to say what the experiment died of." The project lead is left with the unpleasant chore of communicating the issue to senior management.

Storytime

A wizard of a student, having studied deep technical concepts for the last five years in graduate school, is working on her first big project in her new role at a marketing firm. The day before she's supposed to share her results, her manager casually tells her to craft the analysis results into a "story" and reduce the content down to a single PowerPoint slide.

"You have to speak to them like they're in fifth grade," says the manager.

She reluctantly obliges, even though she knows there will be scientists in the audience. She believes the presentation has already been pared down enough. And she's tested the comprehensibility of her delivery with non-technical coworkers.

"Trust me," the manager says, "you don't want this group to ask any questions." So, much of the technical efforts and critical thinking are stripped away as the work is condensed into a simple headline.

At the presentation, most of the audience members nod along while the results are presented. Some wonder, "do we really need fancy 'data scientists' if they're just finding easy answers?"[3] Others, with some technical knowledge, wonder why more technical aspects of the projects have been left out.

The whiz kid reflects on the presentation and feels like a lot of nuance was lost. In some ways, she feels like she has betrayed the original work.

The Telephone Game

At a casual meeting over coffee, a data scientist mentions to her business counterpart an interesting discovery she's found in the company's data. You see, she's in the early stages of the analysis and hasn't had a chance to dig in too deep. But a very cursory review showed that 75% of survey respondents said they would be repeat customers.

After the meeting, the data scientist goes back to her desk and reviews the previous analysis. She once again reviews that 75% statistic and realizes it was a total of 8 people out of hundreds who had responded. After some digging, she learns the question was only recently added to the survey; the question is so new none of the customers who responded to saying they would become a repeat customer have yet to actually become repeat customers.

A month later, at a company all-hands meeting, executives brag about the success of their customer retention efforts. They say 3 out of 4 of their customers are repeat customers based on the responses of hundreds.

The data scientist realizes this factoid came from a casual coffee meeting and was never meant to be presented without vetting. At this point, the factoid has been repeated so many times within the company, it's said as if it were self-evidently true. She wonders how much she could push back on the company using it—and if it's even worth doing so.

Into the Weeds

A data scientist applies the appropriate methods to a challenging problem, and by all accounts answers the business questions. But his resulting presentation to the project team is too technical. Little to no effort was done to tie results back to business value in meaningful way.

---

[3]For more on this perception, check out "Data science done well looks easy - and that is a big problem for data scientists" by Jeff Leek at simplystatistics .org/2015/03/17/data-science-done-well-looks-easy-and-that-is-a-big-problem-for-data-scientists.

His attempts to be recognized as an esteemed technical expert come out as technobabble, and while the stakeholders certainly think the work sounds impressive, they leave the room without clear direction on what do to next. In their mind, because it couldn't be presented in a way that could be understood, the project wasn't actually done yet.

The problem begins to self-perpetuate: the data scientist is asked to go back and work on creating a better solution to finish the project. The data scientist goes further into the weeds . . . .

The Reality Check

A data scientist conducts a market research analysis, but there was no way to execute the solution as a strategy in the market. It's too disconnected from how the business operates. If the company existed in a utopia of endless data quality, budget, and time, it would be a great solution! As it is, however, it's a great solution to an ideal problem—just not the best solution to this problem.

The data scientist, however, is adamant. He wants to implement this the "right" way (aka *his* way). He arrogantly tells his business colleagues they must figure out how to implement it. Finally, a senior partner steps in and says the entire project will be canceled if they can't figure out a path forward on the current trajectory.

"What else can we do?" the senior partner asks. (Until now, nobody had asked this question.)

That's when they, as a team, come up with a way for everyone to win.

The Takeover

After years of interacting with clients in the insurance industry, a project team adds a data scientist to help them analyze years' worth of client data. The data scientist was recently promoted to a senior data scientist, but this "senior" title has gone to their head.

The team has worked hard to establish trusting relationships with each client, but the senior data scientist, overanxious (and perhaps a bit too eager) to fix problems and prove their worth, wants a meeting with the client, saying, "I can't do anything until we have a meeting with the client." Rather than viewing themselves as part of the team, they see themselves as the savior-consultant who will save the company by bringing them into the data space.

While there is nothing like hearing firsthand from the client and getting the opportunity to ask probing Data Head–like questions, the project team feels disrespected by the data scientist. For one, it signals a lack of trust in the team's ability to identify the right context, to uncover the real problem, and to make the link to impact.

Second, it diminishes the difficult work the business side has done to build the relationship that created the trust that resulted in the client being willing to share the need. It signals a disregard (consciously or unconsciously) for the risk of an unnecessary meeting or offending the client in some way.

The Blowhard

A statistician is obsessive about presenting anything but the most technical explanations.

Even though there are others on the team who are just as educated or competent, the Blowhard spends more time arguing about which methodology is the best and picking apart examples online and in textbooks. He openly trash talks the business functions as not being smart enough to understand the work completed (even though they have been at the company much longer than he).

Everyone knows how smart he is, and he relishes his status as a sophist. But he doesn't produce results. Creating a presentation is in truth an agonizing process. And his argumentative style masks an underlying analysis paralysis. Every slide he tries to make feels like he is compromising his beliefs.

As a result, he is a trusted voice when there needs to be a devil's advocate on a project—and that's it. But even asking him for that requires sitting through a tirade of jargon and comparisons to previous projects where he wasn't listened to.

For day-to-day work, his opinions come across as more nuisance than nuance.

DATA PERSONALITIES

Ultimately, the failures within each scene show a lack of empathy and respect for each character's contributions. Take a moment and think about that.

So much of the advice around getting businesses ready for data focuses on investments in technology and training up the workforce of the future. And yet so many failures exist at the level of communication. The problems underlying the scenes were not technology or data problems per se; rather, they arise from conflicting personalities who didn't listen to one another. The good news is that it doesn't have to be this way. There are people out there (readers of this book, we hope) who want to listen and understand what people outside of our roles have to say.

Let's talk about some of the different personalities you'll encounter—and the most important personality—the Data Head.

Data Enthusiasts

Working with data for the first time can be fun and exciting. And the possibilities for businesses to leverage data certainly make people eager to try it out. This is not inherently a bad thing. But some become a bit too enamored with the hype. To them, every new thing is awesome. In their thinking, data can solve every problem. And the very process of being presented with a case study, a result, or a chart is proof that a thorough, scientific analysis was undertaken on the subject. Enthusiasts will often sound like Data Heads on their face and might say things like, "Show me the data." However, they don't ask the hard follow-up questions to separate hype from reality.

When working with data enthusiasts, you should both encourage their love of data while simultaneously reminding them that data can't do the impossible. By encouraging a healthy skepticism of data, you can help them become Data Heads.

Data Cynics

A data cynic's personal experience matters more than data science, statistics, or machine learning. As such, cynics often scoff at data workers' contributions. They view data as, at best, an annoying necessity but prefer to go with their gut. When they don't like results, cynics poke holes—belaboring details that go beyond constructive criticism.

You should take a moment and consider why they might be so cynical. Some of the cynicism might be warranted. Is it because they did not grow up in the age of data? Did they watch other data projects fail? You cannot assume they value data as much as you. When working with them, show empathy by listening to what they value. Speak to those values in your communications and show them you're incorporating their domain expertise into the data solution.

Cynics can become Data Heads over time as they gain more trust and confidence in data, but you must lead them there at their pace.

Data Heads

At their core, Data Heads are skeptics. They aren't skeptical to be annoying. Rather, they're simply employing their data critical-thinking skills. Like the enthusiasts, they advocate for data where useful. Like the cynic, they question what ought to be questioned. But a Data Head's healthy skepticism is informed by technical knowledge and domain expertise and is delivered with empathy.

Becoming a Data Head starts with listening to the entire team. Ultimately, everyone wants to be heard and valued. So, you need to know the obstacles they deal with.

CHAPTER SUMMARY

In this chapter, we reviewed seven scenes that come about as different project personalities interact. Each scene highlighted a communication gap where one of the following happens:

- The business side fails to appreciate the work or challenges of the data side. This theme plays out in the Postmortem, Storytime, and Telephone Game scenes.
- The data side fails to appreciate the work or challenges of the business side. This theme plays out in the Reality Check and Takeover scenes.
- The data side refuses to budge from its technical-only role, either because they were oblivious (the Into the Weeds scene) or arrogant (the Blowhard scene).

Next, we discussed how a Data Head should interact with different data personalities. As a Data Head, you must meet people where they are while also directing them toward an improved understanding of data. This can only happen when you communicate with empathy. In the next chapter, we'll talk about what Data Heads can do to create an environment within their organizations that fosters a better understanding of data.

CHAPTER 15

What's Next?

"One learns from books and example only that certain things can be done. Actual learning requires that you do those things."

—Frank Herbert, American author

To ensure your success, this brief chapter aims to give you what you need as you start the next phase of your learning journey: being a Data Head.

A Data Head is someone who:

- Thinks statistically and understands the role variation plays in their life and decision making.
- Is data literate—speaks intelligently and asks the right questions about the statistics and results they encounter in the workplace.
- Understands what's really going on with machine learning, text analytics, deep learning, and artificial intelligence.
- Avoids common pitfalls when working with and interpreting data.

In other words, a Data Head is someone like you.

To be an effective Data Head, you'll need to become a person who uses data to drive change within your organization. Your authors hope this book has provided enough scenarios for you to consider. But remember, your authors haven't experienced everything, and some of the examples were deliberately trivial to fit into this book. The real world is much more complex.

And using data to change the world sounds good in a book, but it's a lot harder to do in practice.

If you feel motivated by this book, we humbly thank you. But the real work has just begun. These important ideas will have nowhere to go if you don't help us spread the word.

Here are some things you can do:

- Create a Data Head working group at your company.
- Institute regular meetups or lunch-and-learns to dive deeper into the topics we've discussed and learn about those we didn't cover.
- Make a commitment to share your knowledge and help others.

At one point, learning new data concepts was the province of conferences, summits, and workshops. In fact, companies and employees relied on events to help support their learning. And yet, as we conclude this book in January of 2021, 10 months into a global pandemic, we can't help but think this event-driven model is no longer as viable as it once was. Moreover, companies might say on paper they invest in the data learning of their employees, but the reality is that training budgets are only getting smaller.

Born from this reality is that companies have shifted the training burden. Before, they saw new ideas around data as being external—something their staff did not know but needed to bring into the fold. As time goes on, new hires are expected to have already had some experience with data. And being a quick study is now a skill in its own right.

Being a Data Head is to expect this new reality: that much of your learning is going to have to happen off the job (versus on the job). It will happen in books like this, online learning, and certificates. Our world has embraced a cheaper delivery of training, which means that you hold the onus to being informed. No matter where you are in the company hierarchy, you cannot offshore your own personal development to events that happen twice a year. You cannot wait for a stellar keynote to feel motivated. Data will not wait until you want to think critically about it. You must continue learning and be accountable for its trajectory.

You now have the right tools and mindset to be a Data Head. Those who can think, speak, and understand data will be able to cut through the noise, hype, and spin. You don't have to be an industry technology titan to use machine learning and artificial intelligence. In fact, while many of the concepts we presented in this book reflect new technology, the problems they present to businesses aren't new: poor quality data, faulty assumptions, and unrealistic expectations—these issues have been around for decades.

At the same time, the hype and promise of what data can do often pulls attention away from these underlying issues. At the start of the book, we offered a series of data disasters that took place when organizations were not thinking like Data Heads. As the presence of data grows, the risk for such errors increases.

At best, when mistakes happen, they are easy to correct. At worst, they waste money; put people's lives at risk; and reinforce stereotypes baked into the data. As a Data Head, it's up to you to ask the right questions, argue with the data, and even have uncomfortable conversations. With the foundation you've built by reading this book, you are up to the task.

Index